THE URBAN CARIBBEAN IN AN ERA OF GLOBAL CHANGE

The Urban Caribbean in an Era of Global Change

ROBERT B. POTTER
Royal Holloway, University of London

Routledge
Taylor & Francis Group

LONDON AND NEW YORK

First published 2000 by Ashgate Publishing

2 Park Square, Milton Park, Abingdon, Oxon OX14 4RN
711 Third Avenue, New York, NY 10017, USA

Routledge is an imprint of the Taylor & Francis Group, an informa business

First issued in paperback 2016

British Library Cataloguing in Publication Data
Potter, Robert B.
    The urban Caribbean in an era of global change
    1. Urbanization - Caribbean Area 2. Cities and towns -
    Caribbean Area - Growth 3. Caribbean Area - Economic
    conditions - 1945- 4. Caribbean Area - Social conditions -
    1945-
    I. Title
    307.7'6'09729

Library of Congress Catalog Card Number: 00-132579

ISBN 978-0-7546-1139-4 (hbk)
ISBN 978-1-138-27304-7 (pbk)

Transfered to Digital Printing in 2009

Contents

List of Figures

List of Tables

Preface

This is not a textbook on Caribbean cities and urbanisation. Rather, it represents a collection of linked, specialist essays concerning the role of towns and cities in Caribbean societies, along with consideration of some of the connections existing between the processes of globalisation and urban development. Some of the chapters also address issues of sustainability by focussing attention on the environment-development interface in the urban environmental context. In the final analysis, the book makes comment about some of the planning needs of the region over the coming years.

My reasons for producing this book at the present juncture were quite straightforward and serve to elucidate the overall aims of the research monograph. Firstly, despite the significance of urbanisation within the Caribbean, the majority of those who are not familiar with the region tend to think of its constituent territories as agricultural economies, which are now highly dependent on tourism. Statistically, this is a gross-oversimplification, for Caribbean nations are generally highly urbanised. Reflecting this directly, while the literatures on tourism and agricultural development in the Caribbean are extensive, there remains a dearth of literature to hand on urbanisation in the region. Most of the material included in the chapters of this monograph started out as papers in specialist journals, and given the circumstances described above, it seemed worthwhile to bring heavily revised versions of these together in an accessible format.

All of the chapters that started out as specialist articles have been subjected to major revisions and changes. For example, chapters 1 and 3 are both made up of the arguments contained in several original papers and reviews. Additionally, it is worth stressing that several of the pieces first appeared in somewhat inaccessible serials. A case in point is chapter 4, which originally appeared in *Cahiers du Tourisme*, Series C, produced out of the Centre des Hautes Etudes Touristiques, Aix en Provence, France. On the other hand, several chapters and parts of chapters were written specifically for this book, including chapter 2 and chapter 8. It is also necessary to acknowledge here that three of the chapters have been co-written, and this is duly acknowledged on the title pages of the chapters concerned, specifically 4, 7 and 8. However, I should also like to express my warm thanks here to Professor Graham Dann, Dr Sally Lloyd-Evans and Jon Pugh for their collaboration in the production of these chapters. I hope that they will be satisfied with the result, but, of course, I should stress that any errors or omissions remain my responsibility alone.

The other reason for writing this book was the fact that in a recent text it was argued that Caribbean cities are likely to be behaving rather like those that are to be found in Latin America. Specifically, the volume concerned hypothesised that neo-liberal policies and attendant global changes are serving to reduce the size of the largest cities found in territories, in favour of the second city, and that social segregation or separation within cities is declining. These arguments seemed to run directly counter to my general observations in the contemporary Caribbean region, strictly defined to include the small islands as well as the larger island and mainland territories. Accordingly, this stands as one of the major issues addressed in the first part of the monograph.

A number of individuals and organisations have helped in the preparation of this book, not least Kathy Roberts, who once again typeset the entire work, despite her very busy schedule of other tasks. Sue May processed the artwork with care and attention to detail. The team at Ashgate Publishers offered advice and support at a number of junctures and proved to be very efficient and supportive. At various stages, as I carried out the fieldwork on which these chapters are primarily based, I was funded by the Economic and Social Research Council (ESRC), the Nuffield Foundation, the British Academy and the Research and Enterprise Fund of Royal Holloway, University of London. I express my gratitude to all of these for the generous financial support which they provided.

Robert Potter
November 1999

1 Urbanisation and Development in the Caribbean: An Overview

In common with the other nations making up what is referred to as the 'Third World', colonialism and dependent development are critical to an understanding of the Caribbean. In fact, colonial status and dependency can be said to represent the whole history of the Caribbean region (Lowenthal, 1972). The legacy of colonial settlement and subsequent orientation to West European economies is witnessed in a large number of shared socio-economic characteristics, foremost among which are open economies with strong agricultural orientations and a marked tendency toward monoculture. But these forces are just as apparent in the form of dependent urbanisation, the realities of which are vividly expressed in the highly skewed and spatially uneven urban settlement patterns that are to be found throughout the Caribbean region. In this sense, Caribbean countries have always formed part of a globalised system of urban places. The small size of Caribbean countries adds a further twist to these socio-spatial characteristics. Post-war trends towards the development of tourism and manufacturing industry as generators of economic change are held by some to be generally promoting increasing levels of what may be referred to as 'urban bias' in development. These are the principal issues which comprise the focus of this monograph.

Aims and Objectives of the Present Volume

The aim of the present volume is to provide a comprehensive overview of both historical and contemporary facets of urbanisation, economic change and globalisation in the Caribbean region taken as a whole. Although traditionally seen as falling within the hegemony of the United States, in the post-war independence period, variant paths to economic development and territorial planning have been followed in the region. These have ranged from state socialist approaches in Cuba, and at specific junctures and to varying degrees in Grenada, Guyana and Jamaica, through to mixed economies of an avowedly capitalist persuasion.

Presently, there is no up-to-date text covering in a detailed, comprehensive and comparative manner the related issues of urbanisation, globalisation and development paths in this strategic world region. The volume published by Cross (1979) still represents a salient work concerning urbanisation in the Caribbean. Although this remains a useful and interesting text, its scope is strongly sociological and theoretical, and the overall approach thematic, covering theories of urbanisation and dependence, the economic order, population structure, social organisation, race, class, education and politics. However, specifics of global change and detailed consideration of particular territories are not included. A monograph by Hope (1986), although supposedly discussing the entire Commonwealth Caribbean, deals specifically with only the four nations of Barbados, Guyana, Jamaica and Trinidad. In the late 1980s, the present author edited a collection of essays which examined the linked processes of urbanisation, planning and development in the principal territories of the Caribbean region (Potter, 1989a). Although comprehensive both in scope and geographical coverage, this collection is becoming dated, and gives relatively little attention to the attendant processes of globalisation in relation to urban change.

In a recently published edited volume, Portes, Dore-Cabral and Landolt (1997) have endeavoured to examine what they describe as the "Urban Caribbean" in the "transition to the new global economy". Although the title of the book refers to the Caribbean in general, in fact, the volume deals specifically with five countries: two Middle American nations, Costa Rica and Guatemala, plus the larger island Caribbean states of Haiti, Dominican Republic and Jamaica. The book thereby relates to the Caribbean Basin rather than the Caribbean *per se*. Indeed, the study specifically excluded the smaller Caribbean island states which in many senses typify the region, arguing that their size effectively precludes the development of secondary cities. From a geographic viewpoint, this argument may be seen as oversimplistic, ignoring the decentralisation of activities within primate core regions, which appears to be one of the characteristics of the urban Caribbean (see Potter, 1995a). These are the types of current change that receive specific attention in the present research monograph.

The Portes *et al.* (1997) volume was conceived in order to examine the effects of the new international economic context on urbanisation in the Caribbean Basin. The country-specific chapters which make up the book are organised around three closely linked hypotheses, each derived from research carried out in the Latin American urban context. Firstly, it is suggested that the greater the shift from import-substitution industrialisation to export-oriented development, the greater the probability of secondary city growth and the relative decline of urban primacy. Secondly, Portes *et al.* argue that increases in poverty and income differentials as a result of structural adjustment programmes (SAPs) have led to reductions in spatial polarisation within Latin American cities as the outcome of the survival strategies adopted by both the middle- and low-income sectors. This, it is argued, has been brought about by a greater intermingling of classes. Thirdly, again derived from the Latin American urban context, it is stressed that the formal and informal sectors are integral parts of the same urban economy, so that informal sector employment functions only imperfectly as a countercyclical mechanism. This is a truism rather than a working hypothesis, and has long been empirically verified (Santos, 1979).

Perhaps not too surprisingly, evidence from the five nations on the first hypothesis concerning urban primacy is rather mixed. Thus, whilst it is shown that Export Processing Zones have grown rapidly, only in Jamaica has there been a marked reduction in urban primacy. Indeed, the evidence of the volume shows quite clearly that in Haiti and Guatemala, urban primacy has increased significantly, whilst it has shown only very marginal recent reductions in both Costa Rica and the Dominican Republic. The volume is therefore forced to conclude that if export manufacturing and tourism projects are located away from the primate centre, then reduction will occur, otherwise primacy will be exacerbated. Effectively, this is to state the obvious – that government policy is likely to be omnipotent.

In the case of some of the smaller island states of the Caribbean which have followed the export-oriented manufacturing development model, such as Trinidad, Barbados and even St Lucia, tourism and enclave manufacturing have developed on the peripheries of the existing core region. This points to a problem with the Portes *et al.* approach, as the analysis only thinks in terms of what is happening between one or two discrete urban nodes. It does not take into account *concentrated deconcentration* within urban regions, and what exactly is happening at the regional scale. It also ignores the plantopolis and mercantile models of urban development which have recently been specifically revised and developed for the Caribbean region (see chapter 2). The present volume specifically aims to provide this wider and more realistic view on urban processes in the Caribbean region.

Problems of field measurement and definition plague the examination of the second Portes *et al.* generic hypothesis, dealing with relative spatial polarisation within urban areas. Although presenting some insightful commentaries throughout the empirical chapters, work on spatial-social polarisation in the five territories is not supported by precise data. Instead, at best, broad visual/cartographic categorisations of social areas are presented and discussed, with little direct reference to the all important issue of scale. For example, whilst social polarisation might be decreasing at the broad neighbourhood/community level, it may in fact be increasing substantially at the micro-neighbourhood level. The volume has to conclude that the evidence to hand does not support the hypothesis of a uniform "reversal" of social polarisation (itself an ambiguous term given the much cited "reversal" of the industrial city social pattern in the developing world city), although it is argued that a trend in this direction can be discerned.

Finally, the third "hypothesis", concerning the close dovetailing of the formal and informal components of the economy, is something of a truism, as previously stated, but in examining it, the book does serve to present a number of useful and very detailed case studies concerning the nature and operation of the informal sector, for example, artisanal shoemaking in San José, Costa Rica, amber and garment workshops in the Dominican Republic, and the Jamaican small business sector. The lack of a single specific reference in any of the chapters to the seminal work of Santos in respect of the informal sector is surprising. A more general comment can be made in this connection, and this concerns the scant reference made to literature published on the topics dealt with in the Portes volume in developing countries outside Latin America - including the Caribbean region strictly defined.

Whilst the reasons for the strong Latin American slant of the Portes volume can be appreciated, it would have been useful for this to have been reflected directly in the title, in terms of dealing with the Caribbean Basin rather than the Caribbean *per se*. Whilst it may be seen as nothing more than a minor linguistic slip to state that "as in the *rest of Latin America* (sic), Caribbean urbanisation has been studied..." (p.4), there are many people in the Caribbean region who would regard this lumping together of the Caribbean with Latin America as symptomatic of a lack of recognition of the specificity of the region.

The lack of universal application of the three hypotheses selected as themes for the comparative projects does serve to lend useful support to a longstanding, but oft forgotten reality – and one with post-modern connotation – that "even Latin American cities" cannot be understood on the basis of blanket notions about less-developed societies. The need is to examine specific sets of national conditions and processes in the context of global trends and commonalities. Hence, the five case studies presented in Portes *et al.* are ultimately useful, but partly in the iconoclastic sense of showing that the overarching themes selected on the basis of Latin

American cities do not fit in a monotonic manner in the context of these relatively large territories of the Caribbean Basin.

This brings us to a final point. The examination of the complex issues surrounding urbanisation, globalisation and development in the Caribbean, as well as being of prime concern to those working in and on the region itself, should assume wider relevance by providing insights for those concerned with other regions. Perhaps no other area of comparable size within the developing world shows such heterogeneity in its political composition, country size, resource base and overall levels of development, ranging from More Developed Countries (MDCs) such as Trinidad and Tobago, Jamaica, Barbados and Guyana, to Less Developed Countries (LDCs) like Dominica, Haiti, St Vincent and St Lucia. The present volume thereby also stresses a vital, but sometimes neglected theme in geographical development studies, namely that the Third World and its problems are far from homogeneous. Thus, it also emphasises that current planning and development issues – such as development from below, agropolitan development, selective closure, aided self-help housing imperatives and the economic role of the informal sector of the economy – all need examining afresh in different political, socio-economic and cultural settings. Perhaps, nowhere provides such an 'island microcosm' better than the Caribbean region.

A Statistical Overview of Urbanisation in the Caribbean

In common with other regions of the developing world, rapid urbanisation in the Caribbean is a product of the post-1945 period. This is so despite the original establishment of urban centres in the region as points of administrative and commercial control at the outset of the colonial period in the fifteenth and sixteenth centuries (Clarke, 1974). But the large-scale movement of rural populations to urban centres, which has thereby also served to swell rates of natural increase in urban populations, has largely occurred since the late 1940s and 1950s. The twin 'push' of rural poverty and the 'pull' of socio-economic opportunities in the urban arena – both real and perceived – have been effective. Caribbean towns and cities have through time had very little to do with manufacturing activities, but where jobs in the formal sector do exist, they are better paid. The existence of enhanced facilities, both in the public and private sectors, has further acted as a spur to cityward migration.

Interestingly, data published by the United Nations show that the Caribbean region is at present not only considerably more highly urbanised than the Third World in aggregate, but is more highly urbanised than the world as a whole (Table 1.1). This was true in 1960, when somewhat in excess of a third of the total population of the Caribbean region were living in towns, as it is of the projected urbanisation level of 64.6 per cent in the year 2000. Currently, it would appear that just over 60 per cent of all Caribbean denizens reside in towns and cities (Table 1.1). This proportion is projected to rise to over three-quarters by the end of the first quarter of the twenty-first century (Table 1.2).

The impact of cities on the socio-economic landscape of the Caribbean region has been dramatic. Thus, West and Augelli (1976) noted that in 1950, the region had only seven cities with a population over 100,000, three of these being in Cuba. By 1970, the number had risen to at least twelve. In the early to mid-1980s, there were at least twenty-four urban places which had reached the 100,000 population level. Indeed, three located in the Hispanic Caribbean, Havana, Santo Domingo and San Juan, had by that stage well and truly passed the million mark (Potter, 1989). Table 1.3, compiled by the present author from the Demographic Yearbook and other sources, shows the population size of capital cities and those with over 100,000 inhabitants.

Data concerning current levels of urbanisation and rates of urban growth are shown mapped in Figures 1.1 and 1.2. Estimated levels of overall urbanisation – the percentage of population living in places defined as towns and cities – are shown mapped for the year 2005 in Figure 1.1. The general pattern is one of relatively high levels of urbanisation, averaging 67.9 per cent for the Caribbean as a whole in the year 2005. Levels of urbanisation in excess of two-thirds are recorded for the Bahamas, Cayman Islands, Cuba, Dominican Republic, Martinique, Puerto Rico, Trinidad and Tobago and United States Virgin Islands. The data on which these figures are based are listed in Table 1.4.

At the same time, Caribbean states are showing relatively high urban population growth rates, with a Caribbean wide average of 2.14 per cent per annum forecast for the period 1995-2000 (Figure 1.2). Notably, Antigua and Barbuda, Dominican Republic, Haiti, Montserrat (before the volcanic eruption), St Lucia, St Vincent and Guyana exhibit annual rates of increase in excess of 2.75 per cent. Levels of urban primacy are also high in the region, with generally 30-60 per cent of the national population living in the capital city region.

The overall level of urbanisation attributed to the Caribbean region between 1950 and 2025 by United Nations estimates is shown in Figure 1.3. The curve for the Caribbean is shown juxtaposed with those pertaining to the More Developed and Less Developed world regions in aggregate. The convergence of the region's overall level of urbanisation on that for the More Developed regions of the world is clearly apparent

Table 1.1   Total population living in towns and cities and level of urbanisation in the Caribbean, 1960-2000

Date	Total population of the Caribbean living in towns and cities	Percentage of total population living in urban areas	
		Caribbean	World
1960	7.7	38.2	33.9
1970	11.1	45.1	37.5
1980	15.7	52.2	41.3
1990	21.6	58.7	45.9
2000	28.8	64.6	51.3

Source: United Nations (1980)

Table 1.2   Levels of urbanisation in the Caribbean, 2005-2025

Year	Percentage of total population living in urban areas
2005	67.9
2010	70.0
2015	72.1
2020	74.1
2025	75.9

Source: United Nations (1989)

Table 1.3   Population of the principal towns and cities of the Caribbean

Territory	Urban Area	Population ('000s)
Trinidad and Tobago	Port of Spain *	250
Jamaica	Kingston *	671
Cuba	Havana *	2,078
	Santiago de Cuba	397
	Camagüey	279
	Holguín	190
	Santa Clara	191
	Guantánamo	170
	Cíenfuegos	119
	Bayamo	122
	Matanzas	112
Barbados	Bridgetown *	88
Guyana	Georgetown *	188
Eastern Caribbean		
St Lucia	Castries *	56
St Vincent	Kingstown *	33
Dominica	Roseau *	20
Grenada	St George's *	31
Montserrat	Plymouth *	3
Antigua and Barbuda	St John City *	24
St Kitts and Nevis	Basse-Terre *	15
Anguilla	The Valley *	2
British Virgin Islands	Road Town *	4
French West Indies		
Martinique	Fort-de-France *	100
Guadeloupe	Basse-Terre *	15
Belize	Belmopan *	3
	Belize City	55
Hispaniola		
Dominican Republic	Santo Domingo *	1,313
	Santiago de Los Caballeros	279
Haiti	Port-au-Prince *	888
Bahamas	Nassau *	154
Puerto Rico/US Virgin Islands		
Puerto Rico	San Juan	1,816
	Ponce	253
	Bayamón	275
	Caguas	174
	Carolina	148
	Mayagüez	210
US Virgin Islands	Charlotte Amalie	12
Netherlands Antilles	Willemstad *	146

* Capital city

Source: various

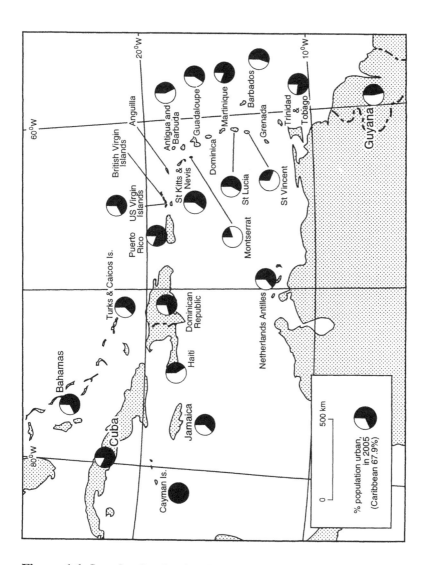

Figure 1.1 Levels of urbanisation in 2005 in the Caribbean

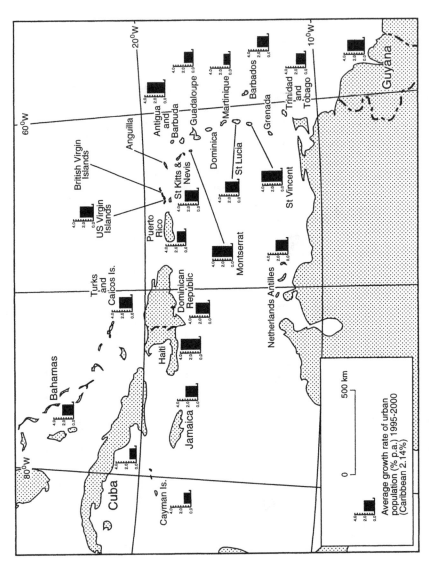

Figure 1.2 Average growth rate of urban populations in the Caribbean 1995-2000

Table 1.4   Level of urbanisation in 2005 and average growth of urban population 1995-2000 for Caribbean territories

Territory	Percentage population urban 2005	Average growth rate of urban population (% pa) 1995-2000
Antigua and Barbuda	42.6	3.51
Bahamas	67.5	2.32
Barbados	54.8	2.33
British Virgin Islands	-	-
Cayman Islands	100.00	1.36
Cuba	81.8	1.31
Dominica	-	-
Dominican Republic	71.1	2.79
Grenada	-	-
Guadeloupe	58.9	1.82
Haiti	41.1	3.90
Jamaica	61.7	2.40
Martinique	81.0	1.31
Montserrat	18.6	4.17
Netherlands Antilles	64.4	2.45
Puerto Rico	80.6	1.71
St Christopher & Nevis	59.7	2.68
St Lucia	57.0	2.83
St Vincent	30.7	4.17
Trinidad & Tobago	77.1	1.94
Turks & Caicos Islands	61.5	2.53
US Virgin Islands	66.5	2.69
Guyana	45.8	3.36
Caribbean	67.9	2.14

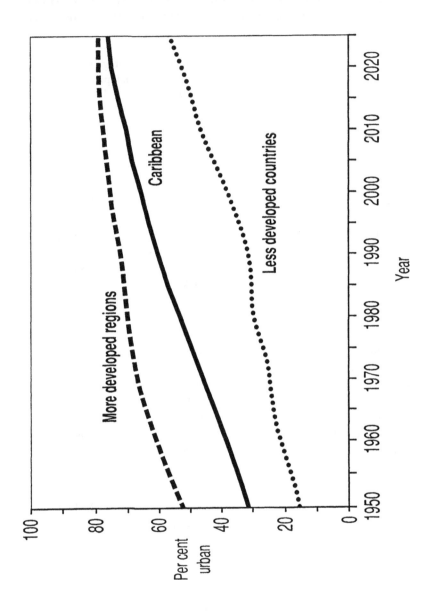

Figure 1.3 **Levels of urbanisation for the Caribbean region in relation to those for the More Developed and Less Developed World regions in aggregate, 1950-2025**

(Figure 1.3). Although this means that the average growth rate of the urban population in the region will fall below 2 per cent per annum after the year 2000, it will still stand at some 1.22 per cent per year by the end of the first quarter of the twenty-first century, as shown in Table 1.5.

The fact that the majority of Caribbean nations show high levels of urban primacy is clearly attested in the region, and as noted, generally, the major town includes well over 40 per cent of the total population of the entire nation. The major growth of primate cities is stressed by West and Augelli (1976) and they note that, generally, the primate urban centre has a population greater than the combined total of the next three largest centres. The pressures on housing, jobs, social services and all manner of facilities that are the outcome of urbanisation over and above industrialisation are commented on by West and Augelli (1976: 120), in the following terms:

> The disproportionate concentration of a territory's wealth, political power, and social services in the chief urban center inevitably shortchanges the small cities and towns. Whether it be government allocations for new schools, roads, and public housing, or the location of new industry or attracting tourists, the smaller urban communities get far less than their share.

The early gateway origins of urban centres within the Caribbean have not only resulted in strong spatial skew at the national level, but also congestion at the intra-urban scale. West and Augelli (1976: 120) note how the chief consideration in the origin of most Caribbean cities was their function as sheltered maritime ports which could be defended with relative ease. Hence, the typical urban site "was on the shore of an embayment protected at the land-ward side by commanding hills on which forts overlooking the sea approaches were built". It is these strategic origins which give many Caribbean capitals their somewhat cramped and huddled character today, with high-density narrow streets which are poorly suited to cars, buses and other commercial vehicles. These commonalities at the intra-urban level are, of course, modified by the prevalent cultural imprint, varying from the grid square iron pattern of Spanish colonial towns to the rectangular and geometrically more varied and *ad hoc* urban form which is more typical of British colonial urban places.

Urbanisation and Development Planning in the Caribbean in an Era of Global Change

The big town - small town, and developed region – underdeveloped region contrasts have given rise to a near universal social and spatial polarity which has set the overall contemporary agenda for planning and

**Table 1.5   Average growth rate of urban population
(per cent per year)**

1950-1955	3.05
1955-1960	3.16
1960-1965	3.92
1965-1970	3.53
1970-1975	3.47
1075-1980	2.80
1980-1985	2.59
1985-1990	2.56
1990-1995	2.40
1995-2000	2.14
2000-2005	1.88
2005-2010	1.66
2010-2015	1.52
2015-2020	1.37
2020-2025	1.22

Source: United Nations (1989)

development in the Caribbean region. It can be argued that the problems involved are rendered particularly acute in small dependent nations such as those that are to be found in the Caribbean (Potter, 1989).

The newly independent formerly colonial territories, perhaps inevitably in seeking to develop, came to equate the state of development with the processes of urbanisation and industrialisation. Having been colonial producers of agricultural staples for so long, when decolonisation afforded political independence, it was perhaps inevitable that it should serve to enhance the desire for a measure of economic independence to go with it (Potter, 1985; Potter and Lloyd-Evans, 1998; Potter, Binns, Elliott and Smith, 1999). The movement toward industry as the "royal road to catching up" (Friedmann and Weaver, 1979: 91) was closely associated with a process of industrialisation by import substitution. Such an approach had, of course, an immediate appeal for countries that had traditionally imported most of their manufactures in return for the export of primary products such as sugar.

However, the call for industry as the chief mechanism for economic development found currency in the West Indies via the arguments of Arthur Lewis (1950), who maintained that an array of industrial activities was possible in the region. Thus started an era of industrialisation by invitation, which boiled down to the attraction of enclave industries – often branch plants established by leading multinationals (Kowalewski, 1982) – by means of affording fiscal incentives, tax-free holidays and infrastructural provision. Such foreign plants frequently produced goods for the overseas home market, manufacturing items like ice hockey equipment, for instance. This policy was successfully pioneered in Puerto Rico, with its 'Operation Boostrap', and was soon followed elsewhere in the Caribbean region (Potter and Lloyd-Evans, 1998), as reviewed in chapter 3.

The relative merits and disadvantages of industrialisation by invitation have been discussed on many occasions, but the arguments are particularly polarised in the case of territories the size of those in the Caribbean. While jobs are created both directly, and indirectly through the operation of income multipliers, there are many problems that are associated with such globalised forms of development. The size and financial might of many First World corporations have undoubtedly meant that they have been able to exact favourable terms in their negotiations with Third World governments. At the extreme, they have, for example, been able to threaten during negotiations, that if their terms are not met, they will, quite simply, go elsewhere (see chapter 3). Equally, even when established, some stay only while the going remains good, leaving during hard times, or when the stipulated tax-free period expires.

Similar arguments apply with respect to that other great component of post-war economic change in the Caribbean, tourism (Bryden, 1993; Patullo, 1997). In this connection, it is generally accepted that there are massive leakages of the revenues that are brought in. The dependency theory school, following the seminal writings of Frank (1969), and locally in the Caribbean those of George Beckford (1972, 1975), maintain that by these mechanisms, the dependent relations of the pre-colonial era have merely been replaced by equally pervasive forms of neo-colonialism (see Clarke, 1986). This is all the more alarming for those who see a geographical corollary in deepening spatial concentration on the primate city/coastal zone, bolstering the kinds of spatial contrasts alluded to previously. This argument is extended into the postmodern context in chapter 4.

In an earlier essay, the present author has attempted to provide an examination of the special relevance of these arguments for small dependent nations such as those found in the Caribbean (Potter, 1989a, 1989c). This discussion culminated in the presentation of a model pointing to the difficulties which confront such nations, and this is reproduced here as Figure 1.4. The exogenous realities for Caribbean territories consist, on the one hand, of the extraction of social surplus value, both by the export of agricultural staples and raw material resources at minimal market prices, and investment and expenditure abroad by élite groups. In the other direction, flows of imported foods, manufactured goods, tourists, enclave industrialisation, new technology, aid and consumerism are characteristic. The geographical pull of accessible and previously well-developed sites with good infrastructural facilities for industry, and of safe and scenic beaches with regard to tourism have served to skew recent developments to those very same leeward coastal tracts that centuries earlier had first attracted mercantile capital (Potter, 1989a). This argument is extended, refined and brought up to date in chapter 2.

The channelling of new developments into the 'relative economic oases' might be said by some to have left an 'economic desert'. The socio-economic contrasts between the two area types are real and affect the daily lives of citizens, but are based on strong urban-rural flows which involve unequal exchange. This may be in the form of low procurement prices for agricultural products or low wages in the tourist and domestic sectors. The disadvantageous terms of trade that affect Third World countries at the global scale are thereby matched by harsh urban-rural terms within countries. This is revealed in long journeys to shops, places of education, health centres, government offices and all manner of other facilities. But the twist in the case of island micro-states is that small territorial size and the historical legacy of concentrated infrastructural provision may well be

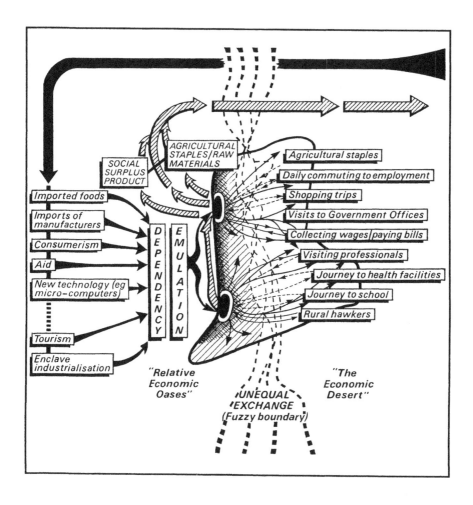

Figure 1.4   A model of Caribbean island development
(Source: Potter, 1989c)

used as a justification for maintaining or even enhancing a highly polarised pattern of spatial development. The model is a greatly simplified one, and will, of course, be modified by territorial size, by culture and, most significantly, by the precise paths to development that countries have pursued over the past twenty year period.

The model of capitalist, centre-out, 'top-down' economic development planning which was applied as an article of faith in the 1960s and 1970s, has been questioned as an invariant path to progress. In its place, ideas for periphery-in and 'bottom-up', collectivist and socialist systems of change and development have been articulated since the 1970s. If countries of the Third World have been held up rather than advanced by their close involvement in the past with the First World, it is logical to posit that they must disengage to some degree if they are to reduce the shackles of their dependence (Potter, 1989a). In the Caribbean, such an approach is, of course, well exemplified by the case of Cuba in the post-1959 revolutionary era (see, for example, Hall, 1981; Susman, 1974, 1987). In Cuba, the main thrusts of planning and development have all involved strenuous efforts to urbanise the countryside and to ruralise the town so as to reduce rural-urban differences in both consumption and production (Acosta and Hardoy, 1973; Gugler, 1980; Slater, 1986). It is precisely these arguments which have underlined the drift of other countries to the socialist model of economic structure within the Caribbean (see Jameson, 1981). In the 1970s, Forbes Burnham and his People's National Congress government declared Guyana a Cooperative Socialist Republic (Payne, 1984). Burnham saw the path that he was taking as somewhat different to that of pure communism, but the approach did, nevertheless, involve state nationalisation of the bauxite and sugar industries. At about the same time, Michael Manley came to power in Jamaica and his People's National Party government stressed equality as the touchstone of its vision of democratic socialism.

But perhaps the most incisive development of the 1970s was the coup led by Maurice Bishop and the New Jewel Movement in Grenada in 1979 (Henfrey, 1984). This brought to power the People's Revolutionary Government after a period of very repressive rule associated with Eric Gairy. As a result of the revolutionary government's concern with nationalism, participatory democracy and its establishment of close links with Cuba, Payne (1984: 20) described the Eastern Caribbean in the wake of these changes as a potential 'sea of splashing dominoes', as other countries showed some signs of drifting toward the left. The struggle between left and right was well and truly once again in Uncle Sam's backyard and would lead to the eventual overthrow and death of Maurice

Bishop, and militarily to the US invasion of Grenada. Not that this brief account of these battles of left and right in the Commonwealth Caribbean represents anything other than a cursory summary of the political economy of development in the region, as the final overthrow of dictators such as Rafael Trujillo in 1961 and Baby Doc Duvalier in 1986 in the Dominican Republic and Haiti respectively bear ample witness (see Payne, 1984; Vendovato, 1986; Potter and Binns, 1988).

However, it does serve to stress that planning the balance which is to be established between urban and rural initiatives and investments is central to the whole issue of development planning. At whatever resolution, planning involves a degree of state control over pure market forces, focusing particularly on the balance that is struck between the autonomy of the individual and the collective responsibility of the state (Conyers, 1982; Cooke, 1983; Potter, 1985; Potter and Conway, 1997). The goals and aims of physical development or spatial land use planning must conform with those of systematic branches of the corporate planning exercise, such as economic and social planning.

Concluding Comments and Outline of Structure

A great deal of popular concern and interest surround the idea that places around the globe are gradually becoming more and more alike. Indeed, superficially there seems to be more than a grain of truth to this observation. Cities and towns, whether they are located in the developed or developing regions of the world, contain the same types of large stores and high-rise apartment blocks, and are increasingly oriented to, if not dominated by, the motor car. Likewise, the consumption of essentially the same range of fast foods, soft drinks, cigarettes and cars, generally manufactured by multi-national companies, is as prevalent in Kingston or San Juan as it is in London or New York.

In more academic terms, writers on urban development and the global economy, such as Armstrong and McGee (1985), have emphasised what they see as the increasing *convergence* or similarity that is occurring between countries with regard to patterns of consumption, as mediated through the dominant capitalist world system. It might be added that this applies particularly to high-income groups in developing countries who in many circumstances lead lives virtually identical to those of their counterparts to be found in cities in the developed world. Urban form seems to be changing in response to such global processes, a theme taken up in chapter 2 of this book. But viewed at the world scale, the reverse seems to be true of patterns of production, with respect to which increasing *divergence* or dissimilarity appears to be characterising countries around the world. This is related to the international division of labour, whereby some countries are regarded as industrial strongholds and others as primary producers. This line of argument is fully articulated in chapter 3.

The role of cities and the nature of the relationships established and maintained between urban and rural areas are crucial in this regard (Potter and Unwin, 1989, 1992). Some maintain that cities are the spatial points at which new forms of production and consumption are introduced, gradually to be spread beneficially to the rural peripheral areas. Others refute this essentially modernisationist argument, pointing to what they see as the role of the city in diffusing patterns of consumption, while production, incomes, wealth and power are ever more concentrated in major urban-industrial nodes, generally those of the metropoles.

This "divergence of production and convergence of consumption" thesis is extremely interesting, for it provides a broad framework for understanding why it is that at the global level, in certain respects, urban form and urban life are increasingly homogeneous, while in other respects, wide differences, as exemplified by shanty towns and massive participation rates in the informal sector, for example, point to contrasts if not of kind, then certainly of degree. How have states assessed and responded to the balance between processes of convergence and divergence, and how have they reacted in their programmes of economic, social and physical development planning, so as to affect the balance existing between urban and rural portions of the national territory? To what extent has the emphasis been placed on increasing domestic production, particularly in the spheres of agriculture and import-substitution industrialisation, as opposed to allowing consumption norms to converge on those of the west? These issues assume a special relevance in small-island micro-states such as those which are to be found in the Caribbean region. The development of tourism is salient in this context, as traced in chapter 4, from the point of view of modernity and postmodernity.

Surprisingly, however, the rise of systematic environmental planning has been relatively slow in the Caribbean (Clarke, 1974). There is an argument that it is the small size, relative insularity and parochialism of West Indian societies that mean that policy makers and politicians rarely think in spatial, geographical and locational terms. Further, this is also reflected in a relative dearth of material on urbanisation and planning issues in the Caribbean, despite one or two interesting essays written on the topic early on (see, for instance, Broom, 1953; Stevens, 1957). Indeed, as Franklin (1979) has noted, with respect to the Third World in general, spatial land use planning is seen by many as far less important than economic development planning, with some effectively regarding it if not as an irrelevance, then certainly as a luxury. Hudson (1986) has stressed the importance of physical planning in the Caribbean region, pointing in particular to the fragile nature of regional ecosystems. The Caribbean is a zone which faces a considerable array of physical environmental hazards, as witnessed by the devastation suffered in Jamaica as a result of Hurricane Gilbert. In addition, the region is associated with periodic economic, political and social upheavals, as demonstrated by events in Grenada and more recently, Haiti. Chapters 5, 6, 7 and 8 specifically focus upon much neglected environmental matters in the Caribbean context, including housing, the informal sector and urban environmental circumstances.

Based on the author's first hand field research, this monograph addresses the twin processes of urbanisation and globalisation as they affect the contemporary Caribbean region. One of the key aims is to focus attention on the fact outlined in the first half of the present chapter, that contrary to popular perceptions, the Caribbean is highly urbanised. In addition, the volume emphasises that the Caribbean region has always been affected by processes of globalisation in respect of its economy, polity and society. The chapters cover pressing topics such as urban change and the evolution of mini-metropolitan regions, the importance of the mercantile and plantopolis frameworks (chapter 2), economic change and the dual processes of global convergence and divergence (chapter 3), tourism, postmodernity and the urban nexus (chapter 4), and the nature of the relationships existing between the state, the informal sector, housing and environmental conditions (chapter 4-7). Chapter 8 provides an overview of Caribbean urban futures with a strong accent on planning. In reality, it is shown that the development of tourism and enclave manufacturing in the Caribbean is leading to new forms of extended urban concentration, and not spatial dispersal *per se.*

2 Towards a Framework for the Consideration of Caribbean Urbanisation and Urban Settlement Systems

The principal theme of this chapter is to exemplify that in considering Caribbean urbanisation, we need to look at the development process itself. In turn, both of these need to be examined in terms of global political economy and global change. We are living in an era of global transformation, but globalisation is not bringing about uniformity, rather it is leading to new and highly specific and differentiated localities and regions (Potter *et al.*, 1999; Potter and Lloyd-Evans, 1998; Klak, 1998). In short, we need to ponder the global causes, and not just the localised effects, of urbanisation. This chapter exemplifies above all else, that in order to achieve this, we have to take a strongly historical perspective. Quite simply, as stressed in chapter 1, globalisation has always been a central facet in the development of the Caribbean region.

Historical Perspectives on the Development of Urban Systems in the Caribbean

The traditional framework derived for explaining the location, size and spacing of urban settlements is referred to as central place theory. The first principles of central place theory were deduced in a comprehensive theoretical manner by a German geographer named Walter Christaller. His work was empirically tested in respect of southern Germany, and his book, *Central Places in Southern Germany*, first appeared in German in 1933, based on his doctoral thesis (see Beavon, 1977; Potter, 1982 chapter 2). The work was translated into English and published in 1966, whereupon it played a pivotal role in the rise of what is still referred to as the "New Geography", which embraced quantification and theoretical sophistication (Christaller, 1966).

Explaining Christaller's central place theory in the simplest terms, it was argued that given environmental uniformity and the operation of supply, demand and the price system in space, the outcome would be a uniform triangular lattice of central places supplying goods and services to the surrounding population. The market areas of these central places (towns and cities) will be hexagonal, and different sizes of central place develop in order to serve different needs. If the demands of marketing goods are paramount, then for every central place of a higher order there will be the equivalent of three places of the next lower order, as shown in Figure 2.1A. If transport efficiency is allowed to dominate the system, then a k=4 pattern emerges, wherein every higher order place is accompanied by the equivalent of four lower order ones (Figure 2.1B). In the so-called k=7 administrative principle, an even larger number of centres is required at each successive lower level of the settlement system (see Figure 2.1C). The outcome of the system is a strict hierarchical organisation of central places by size, with places of the various size orders being evenly spread over the landscape (Figure 2.1).

But empirical observation in many world regions drives home a major point and that is that much of the settlement fabrics of regions and indeed whole continents are strongly coastal in orientation. Another observation is that where settlements extend inland, they frequently show a very strong principle of linearity. It is in this sense that some writers have referred to Christaller's classical central place models as being associated with a bygone age – one of relative insularity – when regions developed from within. It is in this context that the basic central place formulation is described as feudal in its orientation and nature.

In 1970s, an American geographer, Vance (1970) argued that in the seventeenth and eighteenth centuries, mercantile entrepreneurs had to turn outward from Europe because of the long history of parochial trade and what he referred to as "the confining honeycomb of Christaller cells that had grown up with feudalism" (Vance, 1970: 148). With overseas colonialism, merchants were for the first time confronted by an unorganised land mass, whereas in Europe the spatial framework for economic activities was already established in the pre-mercantile period. Vance thereby argued that with the rise of mercantile societies, settlement systems viewed globally started to evolve along more complex lines than previously. The main development, of course, came with colonialism when continued economic growth necessitated greater land resources. Obviously, during this period, ports came to dominate both colonies and colonial powers alike. In colonies, once established, ports acted as gateways to the interior lands. Later evolutionary stages saw increased spatial concentration on the coastal gateways and the establishment of new inland areas for expansion. At the same juncture, the settlement pattern of

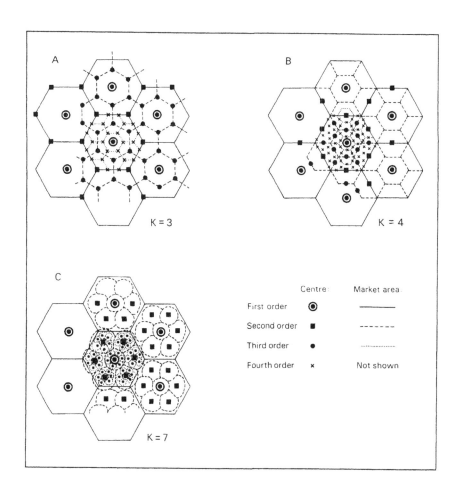

Figure 2.1   Christaller's basic central place settlement systems

the colonial power underwent considerable change, with the original feudal central place pattern experiencing considerable polarisation on the principal ports.

These historical facets of trade articulation led Vance to put forward in graphical terms what he saw as effectively an entirely new model of urban settlement evolution and structure, one that was firmly based on the history of mercantilism and colonialism (Vance, 1970; Potter and Lloyd-Evans, 1998; Potter *et al.*, 1999). Indeed, the highly urbanised nature of the Caribbean territories discussed in chapter 1 has its origins in the mercantile and colonial periods. The colonial gateway and administrative function of Caribbean towns and cities means that Vance's *mercantile model* of settlement location and development is initially highly pertinent. A simplified version of the model is shown in Figure 2.2.

Vance's model suggests that due to the dictates of global political economy, settlement patterns in colonies are far more concentrated than is socially necessary, or perhaps desirable. Indeed, seeing the Vance model as a modification to Christaller's settlement system implies just this. Thus, whilst Christaller stressed the uniform spatial distribution of settlements which would result from supplying a relatively isolated population with essential goods and services, the Vance model shows how this becomes highly distorted and far more concentrated once global trade comes into the equation. Levels of concentration far higher than those envisaged in classical central place theory are characteristic of both the colonial power and the colony, in the form of linear-coastal concentrations and limited inland lineaments, as depicted in Figure 2.2.

The graphical model itself is divided into five stages, each of which depicts the key features of the evolution of globalised urban settlements, in both the colony and the colonial power (Figure 2.2):

- The first stage represents the *initial search phase of mercantilism*. This basically involves the search for economic information on the part of the prospective colonial power.
- The second stage sees the *testing of productivity and the harvest of natural storage*, with the periodic harvesting of staples such as fish, furs and timber. However, no permanent settlement is established in the colony.
- The *planting of settlers who produce staples and consume the manufactures of the home country* represents the third stage. The settlement system of the colony is established via a point of attachment. The developing symbiotic relationship between the colony and the colonial power is witnessed by a sharp reduction in the effective distance separating them. The major port in the homeland becomes pre-eminent.

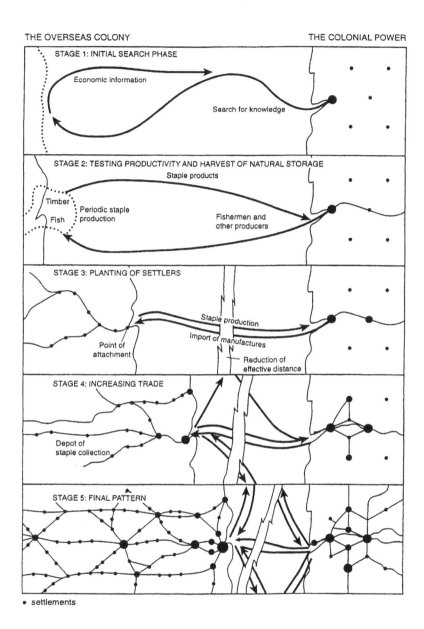

Figure 2.2   A simplified representation of Vance's mercantile model of settlement patterns

- The fourth stage is characterised by the *introduction of internal trade and manufacture in the colony*. At this juncture, penetration occurs inland from the major gateways in the colony, based on staple production. There is rapid growth of manufacturing in the homeland to supply both the overseas and home markets. Ports continue to increase in significance.
- The fifth and final stage sees the *establishment of a mercantile settlement pattern* with central place infilling occurring in the colony; and the emergence of a central place-type settlement system with a mercantile overlay in the homeland.

As already noted, and as graphically depicted in diagrammatic form in Figure 2.2, the hallmark of the mercantile model is the remarkable linearity of settlement patterns, first along coasts (especially in colonies), and secondly along the routes which developed between the coastal points of attachment and the staple-producing interiors. These two alignments are also given direct expression in Taaffe, Morrill and Gould's (1963) model of transport expansion in less developed countries, based on the transport histories of Brazil, Malaya, East Africa, Nigeria and Ghana. But, as is already more than evident, such a framework holds very high explanatory power in the Caribbean context. It certainly fits the realities of the regional settlement fabric better than the relative environmental uniformity which is the characteristic outcome of the operation of classical central place frameworks.

The Caribbean Regional Variant: The Plantopolis Model

In the Caribbean region, however, what may be seen as a local historical variant of the mercantile settlement system can be recognised. This framework specifically acknowledges the foundations of the economy in the globally-oriented plantation system, and maps this into the evolution of the settlement system. This framework is referred to as the Plantopolis model. A simplified representation of this is provided in Figure 2.3. It should be emphasised at the outset that the first two stages of this figure are based on the account provided by Rojas (1989), who thereby originated the plantopolis formulation. On the other hand, the graphical depiction of the sequence and its extension into the "modern era" have been effected by the present author (see Potter, 1995a, 1997; Potter and Lloyd-Evans, 1998).

In Stage 1 of the model, *Plantopolis (1750-)*, it is emphasised that the plantations formed self-contained bases for the settlement pattern, such that only one main town performing trade, service and political control

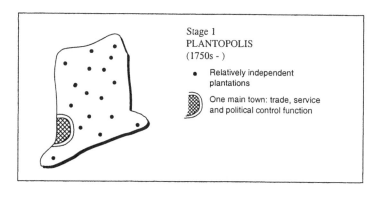

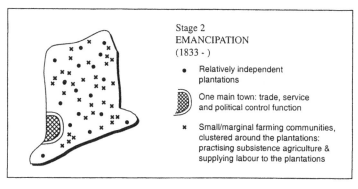

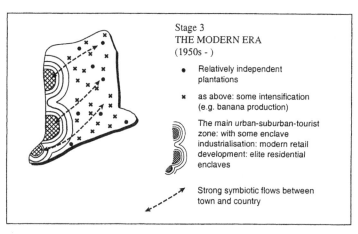

Figure 2.3   From plantopolis to mini-metropolis in the 'modern' era
(Source: Rojas, 1989; Potter, 1995a)

functions was required in addition to the basic agricultural points of control. The model can, of course, be adapted to consider the development of more than one commercial centre if applied to the Greater Antilles. The birth of one main centre and the ascendancy of another subsequently can also be built into the model. Following *Emancipation (1834-)*, small-marginal farming communities, which were clustered around the plantations practising subsistence agriculture and supplying labour to the plantations, added a third layer to the settlement system (Stage 2). The distribution of these, of course, would be modified according to agricultural practice and physical features, especially those of relief.

The primary argument of the present chapter is that the *Modern era post-1950*, has seen an *extension* of this highly polarised pattern of development and change. This is graphically depicted in Stage 3 of Figure 2.3. The emphasis is placed on *extension*, for the changes being outlined may not in all cases amount to a spatial *intensification* per se. In other words, the modern era may well have seen the spatial extension of an increasingly complex concentrated urban region. The outcome has been the development of what may be referred to as *mini-metropolitan regions*.

This has come about largely as the result of industrialisation and tourism being taken as the twin paths to development within a context of increasing similarities in consumption patterns around the world (*global convergence*), and increasing dissimilarity with respect to global patterns of production and corporate ownership (*global divergence*) as noted in chapter 1 (see Armstrong and McGee, 1985; Potter, 1990; and Potter, 1993c in respect of the Caribbean). Stage 3 of the extended model, therefore, shows a situation where the main urban-suburban-tourist zones, replete with limited industrialisation, modern retail forms and elite residential enclaves, stand in marked contrast to the rural areas. The model also serves to emphasise, therefore, that present-day patterns of urbanisation in the Caribbean region are as much based on the development of tourism and related leisure activities, such as golf courses and the like, as it is on housing, manufacturing, retail and other functions.

As shown in Figure 2.3, the complement of the spatial inequality and marked regional differences that are associated with the model, is strong symbiotic flows between the urban and regional zones. It should be stressed once again that outside the smaller insular island states of the Caribbean, the occurrence of this pattern around a number of separate nodes in a single island can just as easily be envisaged. The model, therefore, has wider currency than for the eastern Caribbean alone.

The virtues of the mercantile and plantopolis models are many. Principally, they serve to stress that the development of settlement systems in most developing countries amounts to a form of *dependent urbanisation*.

Certainly we are reminded that the high degree of urban primacy and the littoral orientation of settlement fabrics in Africa, Asia, South America and the Caribbean are all the direct product of colonialism, not accidental happenings or aberrant cases, hence the argument that modernisation surfaces merely chart neo-colonial penetration. According to this framework, ports and other urban settlements became the focus of economic activity and of the social surplus which accrued. A similar but somewhat less overriding spatial concentration also applies to the colonial power. Hence, a pattern of spatially uneven or polarised growth emerged strongly several hundred years ago, with the strengthening of the symbiotic relationship between colony and colonial power. The overarching suggestion is that due to the requirements of the international economy, far greater levels of inequality and spatial concentration of social surplus product are created than may be either socially or morally desirable.

Empirical Verification: The Emergence of Mini-Metropolitan Regions

The third stage of the revised plantopolis model presented here envisages the evolution of compound mini-metropolitan areas. These are the zones of original mercantile urban development which have been stretched out during the so called modern period. In so doing, global capital has seen to it that they have become more and more differentiated in the form of specialised spaces. These principally consist of urban places (central business districts, low-income housing areas, suburban shopping zones), elite residential areas, tourist-oriented zones and areas of manufacturing development. These are the activities noted in Stage 3 of the plantopolis model (Figure 2.3). By this means, it is argued that recent developments have not seen the eradication of sharp rural-urban differences, but rather their replacement with more spatially complex forms of urban development within the urban core area, in a way not envisaged by analysts such as Portes *et al.* (1997). Thus, whilst urban decentralisation and attendant spatial differentiation may be the outcome, little has happened in the way of reductions in gross spatial inequalities in most insular Caribbean territories in the period since the 1950s.

The chapter now turns to consider the historical and contemporary support which exists to substantiate the evolution of complex mini-metropolitan forms of this type in the period since the 1950s. In this instance, recourse is initially made to two Eastern Caribbean territories, Barbados, and St Lucia. Both nations have developed manufacturing activities and tourism in the post-war era. For example, in the post-war period, Barbados developed ten such industrial parks, all of which were strongly urban-suburban in location (see Potter, 1981; Clayton and Potter, 1996, for further details). However, proposals for a peripheral industrial estate did not materialise. St Lucia, however, has focused part of its manufacturing efforts on the second urban node of Vieux Fort. Such developments have, therefore, tended by and large, to concentrate on the pre-existing core urban areas, where the infrastructure was already to be found, and have rarely been used for the purpose of promoting decentralisation outside the existing coastal urban zone. The path to development has largely continued to be top-down.

In the Caribbean region, the promotion of tourism can also be seen as part of the process of globalisation, as territories sell their natural beauty and climatic regimes. This has obviously focused development and change once more on coastal areas, and in a large number of cases, has seen further development of the pre-existing coastal urban zone. This has certainly been so in Barbados, St Lucia and Grenada. There have, however, been some efforts to spread development in the case of St Lucia, to the Soufriere and Vieux Fort urban zones. Spatially, the further development of the primate city region and development in the modern era has led to what may be referred to as compound urban-cum-industrial-cum tourist zones. This is the essence of what is referred to here as complex mini-metropolitan areas.

The Genesis of Mini-Metropolitan Zones: The Empirical Case of Barbados

We can now look at the overall outcome in respect of two territories of the eastern Caribbean, Barbados and St Lucia. The existence of strong spatial inequalities in the case of Barbados is well-attested, notwithstanding the small size of the nation, its rapid development to middle-income status since 1950s, and the ubiquity of its public transport system. Potter (1987) has looked at the development of inequalities in Barbados, both historically and contemporaneously. The first European settlement, Jamestown, was established half-way along the sheltered leeward or west coast, at what is today known as Holetown. Shortly after, rival settlements were established at Bridgetown, Speightstown and Oistins. Over time, despite the early ascendancy of Speightstown as a sugar port in the north, the locational advantages of Bridgetown became pre-eminent. These advantages included the best natural habour at Carlisle Bay, as well as a coastal inlet known as the Constitution River, an area of low terrain inland and good access to the agricultural interior.

The major feature of the development of the urban settlement pattern of Barbados in the colonial period was its remarkable persistence and durability. Indeed, the four major settlements remain as the contemporary urban foci, forming a continuous linear urban corridor. Slowly, this corridor has become differentiated into a complex and functionally-interrelated metropolitan zone. Bridgetown acts as the centre of political and economic control. Data presented by Potter and Wilson (1989) show that since 1940, the shift from sugar production toward tourism and services, and to a lesser extent manufacturing, has seen a continued movement of population to the suburban and urbanised belt (Potter and Wilson, 1989). Some 70 per cent of manufacturing jobs are to be found in the Greater Bridgetown area. If the national distribution of lawyers, banks and general practitioners is scrutinised in detail, then their paucity in non-metropolitan Barbados is clearly evident. Thus, whilst in 1989, 76 lawyers were to be found in Central Bridgetown and 77 in Greater Bridgetown, none existed in the rest of Barbados. Likewise, whilst 147 medical doctors were to be found in Central and Greater Bridgetown, only 19 were to be found in the rest. The distribution of banks between central Bridgetown, Greater Bridgetown and the rest of Barbados was 17, 32 and 16 respectively. Similarly, Potter and Dann (1990) show that 78 per cent of all retail floorspace is to be found in Bridgetown, including all of the recent supermarket-based developments. Indeed, a staggering 97 per cent of all retail floorspace is to be found within the linear urban corridor. The concentration of manufacturing activities has already been exemplified. In the twentieth century, Oistins has become strongly associated with suburbanisation towards Christ Church, and Holetown with high-status west coast tourism, which since 1980 has extended toward Speightstown. All of these developments have given rise to a complex functionally-integrated coastal metropolitan zone.

The outcome can be seen in the contemporary socio-economic pattern, as recently analysed by Potter, Jacyno and Lloyd (1995). An array of 34 variables was taken at the parish level for the 1990 Census, covering a range of housing, utility, amenity, socio-economic, demographic and ethnic dimensions. The data were factor analysed, and the first derived factor was clearly a measure of *amenities and modernity* (Table 2.1). When the scores of the parishes on this factor were mapped, St James was affirmed as the most affluent parish, followed by St Philip, Christ Church, then St George and St Thomas (Figure 2.4). A somewhat lower amenity score was recorded for the Greater Bridgetown area, and substantially lower values pertained to the rural areas. Factor 2 emerged as an overall measure of *urbanity*. High positive scores pick out the coastal belt of St Michael, Christ Church and St James, in that order. Factor 3 represented a measure of *rurality*, and Factor 4 *youthful areas* (Figure 2.4).

Table 2.1   **Factor analysis of socio-economic, demographic and housing data from the 1990/91 Census for the parishes of Barbados**

Factor	Original variable	Loading	Interpretation
1	Houses pre-1970	-0.96	"Amenity/Modernity"
	Car ownership	+0.86	
	Population over 60	-0.86	
	Solar water heating	+0.84	
	All facilities available	+0.83	
	Houses with 7+ rooms	+0.83	
	Wall/concrete houses	+0.80	
	Washing machines	+0.79	
	Wood walls	-0.73	
2	Gas for cooking	-0.97	"Urbanity"
	Rented housing	+0.96	
	East Indian	+0.93	
	Ownership	-0.91	
3	Wood/charcoal for cooking	+0.75	"Rurality"
	Public standpipe	+0.73	
4	Population under 5	+0.91	"Youthful area"
	Population under 15	+0.80	

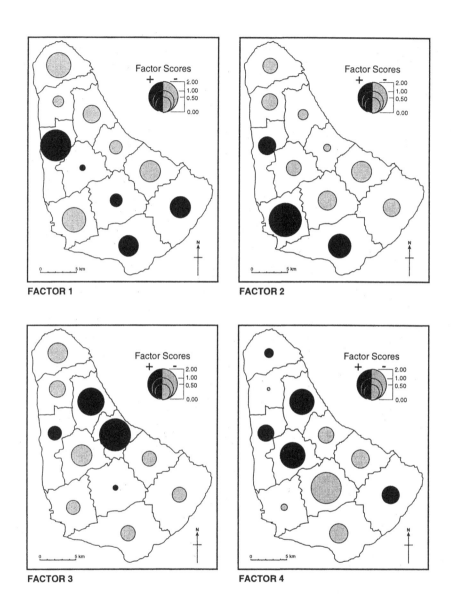

Figure 2.4   Factor scores for the parishes of Barbados

If finally, the scores achieved by the eleven administrative parishes on the first two factors is shown in the form of a graph, a clear typology of socio-economic areas within Barbados is derived (Figure 2.5). The principal urban core of St Michael occupies the bottom left portion of the graph. The affluent coastal parishes of St James and Christ Church are identified together at the top right of the graph. In this way, the extended mini-metropolitan zone of contemporary Barbados is clearly identified by this analysis. In contrast, the top left segment of the graph brings together the three socially upwardly mobile suburbanising parishes of St George, St Thomas and St Philip. Lastly, the rural northeastern parishes of St Andrew, St Joseph, St Peter, St Lucy and St John together make up the bottom left of the graph. The pattern is shown mapped in Figure 2.6.

Taken in aggregate, these results show that a clear typology of areas by socio-economic and demographic traits characterises Barbados. This represents an intensification of the historical pattern of inequality which dates back to the mercantile period and the development of what may be referred to as a plantopolis-oriented pattern of settlement and economic activity (see Potter, 1995a). Hence, the strong leeward coastal urban-suburban-tourist zone, replete with enclave manufacturing, modern retail forms and elite residential enclaves, stands in marked contrast to the rural areas. However, between these two long-established area types exists St Philip, St George and St Thomas as socially upwardly mobile parishes which are actively being drawn into the suburban ambit. Hence, the analysis pinpoints very clearly, both the components of inertia and change which characterise the contemporary socio-economic profile of Barbados, as a middle-income developing nation.

The Genesis of Mini-Metropolitan Zones: The Empirical Case of St Lucia

We turn now to the case of St Lucia, the original settlement of which owed much to defensibility (Potter, 1985; Hudson, 1989). Castries is located on a protected inlet with high land all around, and the town has a long tradition of military association. The town's physical expansion has been blocked other than in three directions – inland to the east along the Castries River, along the narrow coastal strip to the south of the harbour, and north/ northeast toward the northern beaches (Sahr, 1998). The area to the north of Castries has become particularly associated with the development of tourism and recreation. As an illustration, of the fourteen resort hotels featured in the British Airways 1996 Caribbean brochure, eleven are located to the north of Castries, one just to the south, and two around

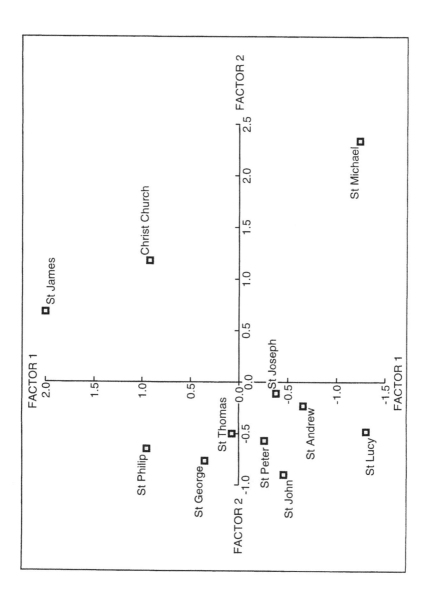

Figure 2.5 Cross-classification of the scores of the parishes of Barbados on the first two factors

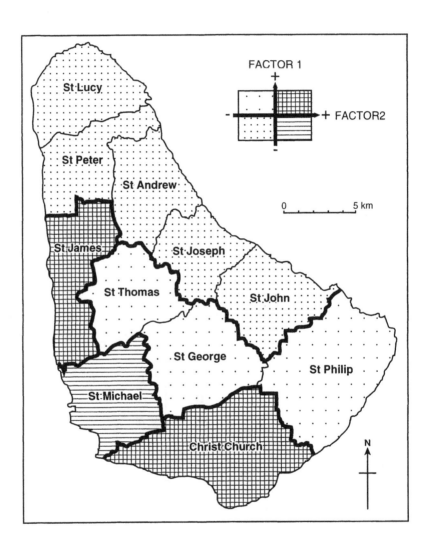

Figure 2.6   A typology of socio-economic areas in Barbados

Soufriere. The Rodney Bay complex, consisting of hotels, restaurants, shops and a marina, was developed in the 1960s, but has been criticised on environmental grounds. High income residential areas also exist to the north, such as Cap Estate. Meanwhile, the improvement of tourism-related facilities has been ongoing apace in Castries, especially the Pointe Seraphine duty free area, and the zone around the central market.

Castries, as the main port and second (regional) airport, remains an important manufacturing area. South of the capital, at Cul de Sac, the United States-based multinational Amerada Hess Corporation entered into an agreement with the Government of St Lucia for a petroleum storage facility and superport in 1977. Vieux Fort, in the south is the site of the country's main international airport at Hewanorra, and the second seaport, and in the early 1970s, was designated as the chief industrial development area. This facility consisted of a 100-acre industrial estate, situated within two miles of Hewannora International Airport and the Vieux Fort dock.

The outcome is shown in the analysis of socio-economic data taken from the 1991 Census of St Lucia. A set of thirty-four variables, essentially similar to those selected for Barbados, was subjected to factor analysis (Table 2.2). The first factor derived, much as in the case of Barbados, is a measure of overall *amenity/affluence*. When mapped, Castries metropolitan, Castries suburbs and Castries rural all recorded positive scores, as did Gros Islet to the north (Figure 2.7). The remaining areas of the national space exhibit negative scores, although that pertaining to the southern secondary town of Vieux Fort is close to zero. The second factor appears to reflect the existence of *youthful areas*, and pinpoints the City area, together with the three western quarters of Canaries, Soufriere and Choiseul. Factor 3, a measure of *modern housing*, exhibits a complex pattern, whilst Factor 4 stresses the occurrence of *squatters* in metropolitan Castries, Vieux Fort, Laborie, Micoud and Dennery (Figure 2.7). When Factor 1 is cross classified with Factor 2, the salience of the coastal metropolitan zone is cogently highlighted, grouping together Castries suburban with Castries rural and Gros Islet (Figure 2.8). At the core of this complex urban coastal zone is Castries metropolitan, which scores negatively on Factor 2 (youthful areas). Taken together these areas are clearly differentiated from the non-metropolitan zones of St Lucia. The basic divide between the western and eastern parishes is that those to the east are generally poorer and more youthful than those to the west. Once again, the data attest to the existence of a sharply polarised map of development, and one which runs strongly in favour of the long-established urban-coastal metropolitan zone, as shown in Figure 2.9.

Table 2.2 **Factor analysis of socio-economic, demographic and housing data from the 1990/91 Census for the quarters of St Lucia**

Factor	Original variable	Loading	Interpretation
1	Television	+0.98	"Amenity/Affluence"
	Electrical light	+0.98	
	Telephone	+0.93	
	Outside kitchen	-0.93	
	Gas for cooking	+0.89	
	Charcoal for cooking	-0.89	
	Video recorders	+0.85	
	Gross earnings	+0.76	
	Radio	+0.75	
2	Housing pre-1970	-0.92	"Youthful areas"
	Housing post-1980	+0.90	
	Population over 60	-0.90	
	Population over 85	-0.89	
3	Average number of rooms	+0.87	"Modern housing"
	Wall/wood houses	-0.86	
	Wall/concrete houses	+0.73	
	Disabled population	-0.67	
	Standpipe	-0.65	
4	Squatted housing	+0.86	"Squatting"
	Water piped to yard	+0.68	

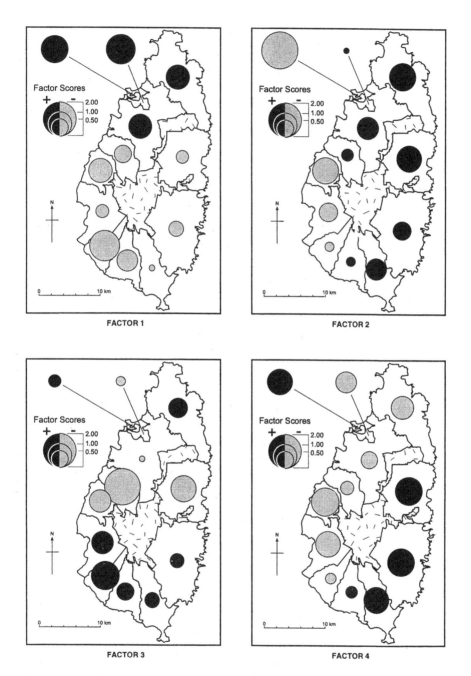

Figure 2.7   Factor scores for the quarters of St Lucia

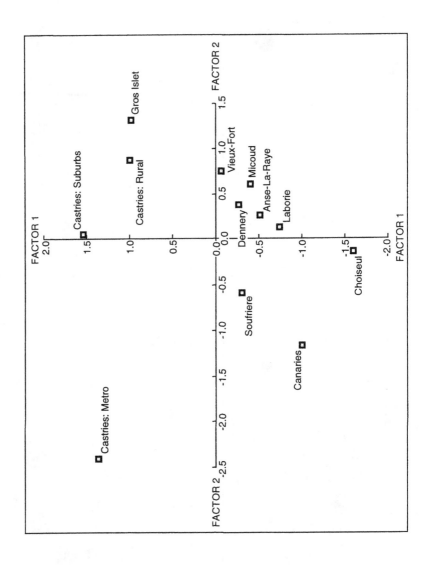

Figure 2.8   Cross-classification of the scores of the quarters of St Lucia on the first two factors

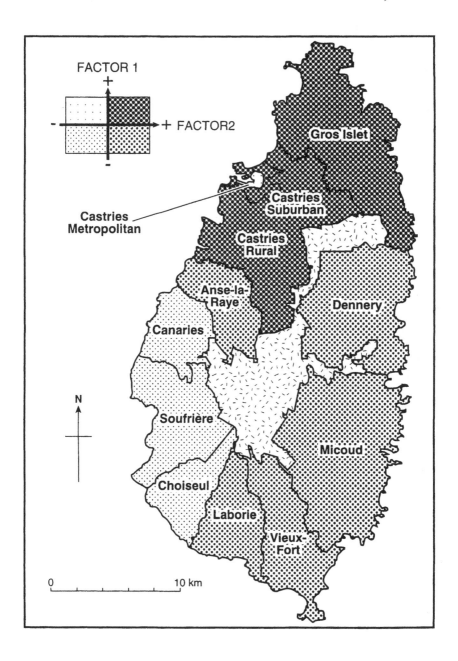

Figure 2.9 A typology of socio-economic areas in St Lucia

The Genesis of Mini-Metropolitan Zones: The Case of the Port of Spain Urban Corridor, Trinidad

The two case studies provided previously examined broad census data to reveal clear socio-economic spatial differentiation. This third and last example looks at Port-of-Spain, Trinidad from a more discursive and historical point of view.

Port of Spain stands as one of the largest urban regions in the Commonwealth Caribbean. When considering its population size, a figure of 250,000 is currently quoted for the wider urban region, making it second only to Kingston's metropolitan region in Jamaica, which has around 671,000 inhabitants (see Table 1.3). In 1992, the combined population of Port of Spain, together with the surrounding county of St George was over 500,000. With a population of 1,169,000 for Trinidad as a whole in 1993, this means that just under half the population of the island lives in the capital city.

It might be argued that Port of Spain is typical of Caribbean urban areas, in that it has grown with little planned influence or guidance, and that it is the outcome of 'uncontrolled urbanisation' (Conway, 1989). On the other hand, it can be argued that in certain respects its growth and development display features of the ways in which Caribbean cities are likely to evolve in the near future, with a tendency toward the distinct zonation of urban land uses.

Looking at a map showing the concentration of settlement and population in Trinidad reveals a typical Caribbean urban pattern: the strong coastal alignment of settlements on the sheltered or leeward coast (Figure 2.10). This spatial concentration can be directly related to the mercantile model of settlement growth, where the urban areas have grown at the points of interaction between the colony and the colonial power. Hence, settlement is both more linear and more coastal than is predicted by Christaller's classic central place model. However, Port of Spain was not the initial principal urban settlement of the evolving colony of Trinidad. St Joseph was the original capital following Spanish colonisation in 1592. It was eventually abandoned, and in 1757, a new Spanish Governor arrived and relocated his residence to Port of Spain. At that time, the settlement was little more than a fishing village and a port of call for ships trading in tobacco, its population standing at less than 400. Port of Spain grew with the development of the plantation economy, becoming the chief administrative and commercial centre. By 1784, its population had increased to over 1,000 then to 4,000 by 1797, when it had become one of the busiest towns in the Caribbean, accommodating French settlers, British merchants and Spanish administrators. From that point onward, Port of

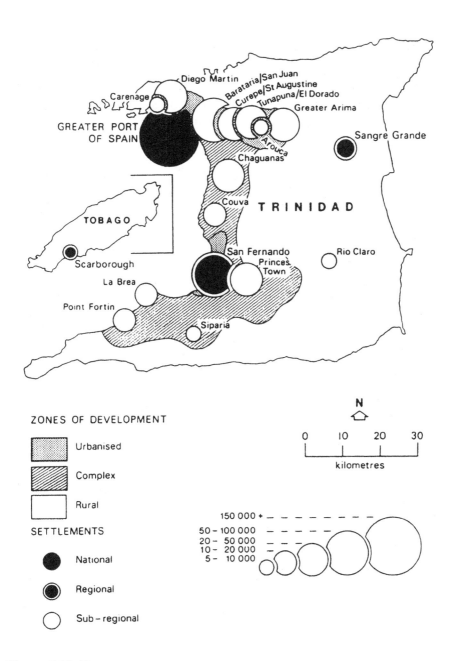

Figure 2.10 Zones of development in Trinidad and Tobago, showing high levels of urban concentration

Spain retained its urban primacy, despite the growth of San Fernando as the southern focus of growth and development, centred on the 1911 oil boom.

Port of Spain flourished in the nineteenth century. By 1838, the city population had grown to 11,701. As a port with a good location, it was ideally suited for British mercantilist efforts in the Americas. The core of Port of Spain, built before 1790, developed on a north-south grid pattern which forms the present-day Central Business District (CBD). The site of the city is constricted to the north by the mountains and to the south by the sea, so that growth has occurred in a linear fashion. To the east of the city centre, poorer urbanites and in-migrants to the city made their homes in the nineteenth century so that high-density residential areas and yards were constructed in the East Dry River and Laventille areas. The uncontrolled and piecemeal development of former small coffee and cocoa estates, many of which had been abandoned, has given rise to a clearly identifiable low-income sector of the city. In-migration and cityward migration in the twentieth century have served to increase densities in these eastern districts of the inner city.

Meanwhile, the better off started to settle to the west of the CBD. The area of Woodbrook was developed for upper-class suburbanites. From 1838, improvements in transport and communications, such as a mule-drawn tramway and horse-drawn carts, allowed further suburban expansion. This continued through to the twentieth century, with British and European families distancing themselves from the east of the city. St Clair farther to the north gradually overtook Woodbrook as the premier status address. The westward expansion of the high-status area continued. But later, elite residential development was diverted northwards into the Maraval, St Ann's and Cascade Valleys. Thus, the growth of Port of Spain basically followed a sectoral pattern. The high status zone developed to the west and to the northwest of the CBD, and the low status area to the east. The planned and controlled expansion of the northwestern high-status sector stood in marked contrast to the development of the east.

Turning back to the regional context, urban concentration and urban primacy in Trinidad increased up to 1946, but thereafter the process of movement of population to the suburbs started to become very important. The conurbation spread in a linear fashion, so that the St George county acquired a dominance which it has never lost. By 1960, Port of Spain had reached a population of 94,000, but from that time onwards, the central city or city proper started to lose residential population.

Urban growth occurred from Port of Spain toward the east incorporating St Juan, Curepe, St Augustine, Tunapuna, Arouca, as far east as Arima (Figure 2.10). Thus evolved the Capital Region, or east-west linear urban corridor. This area now has the best facilities available within the nation. It is served with pipe-borne water supplies, and a sewerage system covers the entire central urban area. By 1970, the Capital Region was served by 100 miles of major urban roads, including two major east-west expressways, the Western and Eastern Main Roads. The former Eastern Main Road light railway has also been converted into a Priority Bus Route, giving fast access to Port of Spain from the west and out again when motor traffic is clogged at peak flows.

The Capital Region is the primary locus of employment opportunities within Trinidad, especially in the manufacture and service sectors. Thus, whilst 32 per cent of the national working population reside in the Capital Region, it provides 53 per cent of total national job opportunities. Some 78 per cent of the value of the output of the manufacturing sectors of Trinidad and Tobago is produced in the county of St George. In addition, it is estimated that the Capital Region is home to 63 per cent of national jobs in the service sector. However, it is likely that jobs in the informal sector, those activities which go largely unrecorded, such as street selling through to small scale manufacture, make this total much higher.

As well as the gradual movement of population from the west to the east of the Capital Region, and the daily movement of commuters from the east to the west and back again, Port of Spain is also the focus of an undisclosed number of commuters who travel across the county boundaries into the region to work. Figures suggest that as many as 15,000 travel in each day from Caroni, 5,000 from St Andrew and 4,000 from Victoria. Such high levels of development and movement mean that the Capital Region faces more than its fair share of environmental problems, including serious traffic congestion on the east-west arteries, and destruction of nearby forests and problems of water conservation.

Conclusion

A central issue is whether such concentrated patterns of urban-based development are environmentally sustainable in small island ecosystems. This is a particularly salient question given the central role which is being played by tourism in most Caribbean states in the post-war period. It is the process of global development that is giving rise to the distinctive locality-based nature of urbanisation and urban development in the Caribbean region, as elsewhere in the Third World. Thomas (1974, 1988) has addressed some of the issues involved in relation to the Caribbean. Thomas argues that there has been a profound and persistent divergence and inequality between resource use on the one hand, and domestic demand plus basic needs on the other. The same author has argued that what is needed is a countervailing convergence or reconciliation between domestic resource use and social needs, by means of the imposition of consumption planning and control. In the present context, this would appear to suggest the need for a thorough evaluation of the efficacy of strongly polarised patterns of development, based as they are on the evolution of complex interrelated mini-metropolitan functional regions within the majority of Caribbean territories. There is the suggestion that the urban developments which have occurred in the third, or modern era of the plantopolis model are not likely to be sustainable in the long run, when considered both socially and environmentally. Linked directly to these issues, in the next chapter, two of the principal processes of global change, referred to as convergence and divergence, are fully explored in relation to the Caribbean urban condition.

3 Global Change and Caribbean Urbanisation: Global Convergence and Global Divergence

Introduction

Just as historically the mercantile-colonial period saw the development of highly skewed patterns of settlement in the form of what has been be referred to as 'plantopolis', as shown in Chapter 2, so post-war developments in Caribbean economy, society and polity have witnessed still greater concentration on coastal-linear zones. In this chapter, this argument is elaborated with regard to developments in the agricultural, manufacturing, tourist and retail-commercial sectors of the economy. It is in the light of highly restricted production possibilities that these aspects of Caribbean urban development must be interpreted. Specifically, enclave manufacturing, offshore data processing, parachute tourism, informal sector employment and housing can all be seen as direct reflections of the process of increasing *global divergence*, which is attributable to the latest stages of the International Division of Labour. These facets mean that despite the diminutive size of Caribbean island micro-states, strong parallels can profitably be drawn between them and other Third World territories.

On the other hand, in respect of both individual and collective consumption, there are signs of increasing *convergence* on western norms. This is witnessed in trends such as suburbanisation and the leapfrog decentralisation of elite groups, the "Miamization" of tourist resorts, postmodern architecture, speculative land holding, rapidly rising land prices, westernised dietary preferences and many aspects of urban morphology and social structuring. It is in these respects that the hegemony of North America is most cogently revealed in the contemporary globally-oriented Caribbean context.

It is argued in the present chapter that such uneven development in the Caribbean region has been intensified in the period since 1945 as a result of

49

the operation of what were referred to in chapter 1 as the dual processes of global *'convergence'* and global *'divergence'*. The divergence convergence debate owes much to Armstrong and McGee (1985) and their analysis of Asian and Latin American urbanisation.

Starting from the observation that during the 1970s a series of fundamental changes occurred in the global economic system, not least the slowdown in the major capitalist economies and rapidly escalating oil prices, Armstrong and McGee stressed that such changes have had a notable effect on urban systems in Third World nations. Foremost among these influences have been changes in technological processes in some industries which have allowed the dispersion of manufacturing industries to low-labour cost Third World territories, and the increasing control of trade and investment by transnational corporations. It is the former trend which has seen the establishment of Fordist (assembly-line) production systems in the Newly Industrialising Countries (NICs) such as Taiwan, whilst smaller scale and more responsive, or so called 'flexible' systems of production and accumulation have become more typical of western industrial nations (see Wallace, 1990). It is argued that productive capacity is thereby being channelled more and more into selected hands and a limited number of urban nodes. It is necessary, therefore, to distinguish between export-oriented industrialising countries, large inward-looking industrialising countries, raw material exporters and low-income agricultural exporters (Armstrong and McGee, 1985; see also Potter, 1990). Thus, increasing division of labour and the increasing salience of large corporations is resulting in enhanced heterogeneity or *divergence* between nations with respect to patterns of capital accumulation and productive capabilities.

Some commentators have, therefore, pointed to the increasing similarity which appears to characterise urbanisation in Third World and western nations (Browning and Roberts, 1980). However, Armstrong and McGee observe that whilst this may generally be true of major urban centres, it is hard to accept the veracity of this argument across the Third World taken as a whole, or indeed, for all geographical areas within particular territories. They continue by arguing that there appears to be at least one respect in which a predominant pattern of what may be referred to as *global convergence* is occurring. This is in the sphere of consumer preferences and habits. Thus, Third World cities may be seen as prime channels for the introduction of imitative lifestyles, many aspects of which are sustained by imports or the internal activities of transnational branch plants. This in turn is linked to patterns of indebtedness, massive collective consumption and increasing inequalities (Armstrong and McGee, 1985). Such disparities reflect the fact that it is the members of the upper-income elite group that are most able to adopt and sustain a wide variety of so called 'modern' consumer habits, although the middle- and indeed the lower-income groups are increasingly coming to share in these aspirations and/or lifestyles.

Patterns of Global Divergence and Contemporary Caribbean Urbanisation

It was noted above that there appears to be a growing divergence or dissimilarity that is characterising countries with regard to their patterns of economic production. This is related to the international division of labour, whereby countries have increasingly come to occupy different productive niches. Thus, whilst some nations have, since the Industrial Revolution, become well-established industrial strongholds, a few newly industrialised countries, and others have been held as primary staple producers. Yet others have adapted to the demands of international tourism with the production of acceptable places and spaces for tourists and tourism. Clearly these processes relating to global divergence are germane to an understanding of urbanisation and change in the contemporary Caribbean. Perhaps most tempting is the argument that post-1947 trends in the development of manufacturing, offshore data processing and tourism have led to increasing urban bias and uneven development.

The newly independent formerly colonial territories, perhaps inevitably in seeking to 'develop', came to equate this with the processes of urbanisation and industrialisation in the form of 'a royal road to catching up' (Friedmann and Weaver, 1979). Thus, just subsequent to the period of the granting of independence, Downes (1980) noted that the less developed countries of the Windward and Leeward Islands had generally followed in a passive and accepting manner, the progression of approaches to development which had earlier been pursued by their more developed neighbours such as Jamaica, Trinidad, Barbados and Guyana, who in turn had followed the role models provided by the colonising power. Having been colonial producers of agricultural staples for so long, when decolonisation afforded political independence, it was perhaps inevitable that it should serve to enhance the desire for a measure of economic independence to go with it. Given their peripheral dependent capitalist status, the conventional wisdoms followed were those of export-led monoculture, industrialisation by import-substitution and the invitation to enclave enterprises.

Thus, relatively little attention has been given to agriculture for domestic purposes. For centuries, agriculture has dominated the economies of these islands, frequently based on the plantation system and high levels of foreign ownership. The plantation has been associated with a pattern of *persistent poverty* by the Jamaican social scientist George Beckford (1972). In some areas, since emancipation there has been a quite substantial growth of the smallholder sector (Shepherd, 1947; Rojas, 1984). Indeed, in the Windward and Leeward Islands, the eclipse of the former plantation produced staple sugar, and the introduction of the replacement staple bananas, has given rise to a series of land redistribution schemes (Rojas and Meganck, 1987).

However, both the large- and small-scale sectors have continued mainly to contribute to export staple production since independence. Agricultural productivity has decreased for the region as a whole, and many countries are now heavily dependent on food imports (Gumbs, 1981). Long established systems of land tenure such as those based on the French system of equal inheritance in St Lucia, and continuing export crop monoculture, are seen by many as causal. Other factors include ongoing processes of soil erosion and land degradation. Thomson (1987) has cogently explored what he sees as the role of the new staple of the eastern Caribbean, the banana. He refers to the crop as 'Green Gold' and argues that in a good year, the crop can be a veritable licence to print money for the British company Geest, who are the sole importers and marketers of the crop into Britain. Yet, for the individual small farmer, production is an extremely precarious business, not least because of the devastation that can be wrecked by hurricanes, such as Allen in 1980. Thus, poverty and dependence continue long after political independence has been achieved. These issues are, of course, likely to be intensified in the wake of the unification of the European market and the ending of preferential trade agreements with non-EU countries.

Despite the limited attempts made by some governments to increase agricultural diversification, Barbados among them, the basic nutritional needs of the population are frequently either being met by the costly import of foodstuffs, or not at all. Thus, data show that for many Caribbean territories, food imports amount to around 30 per cent of all imports. Items such as tinned meats and fish, rice, biscuits and flour are typically imported. There is clearly pressing need for the introduction of more efficient modes of using agricultural land, along lines which will encourage self-sufficiency and sustainability.

Import substitution as a means of development was always likely to hold a special appeal for the Caribbean, given its position as an exporter of primary products. As noted in chapter 1, the call for industry as the basic mechanism for economic development found a special currency in the West Indies, via the arguments of Sir Arthur Lewis (1950), who maintained that a wide array of industrial activities were suitable in the Caribbean.

Thus started an era of industrialisation by invitation, which boiled down to the attraction of enclave industries – often branch plants established by leading multinational companies (see Kowalewski, 1982) – by means of affording fiscal incentives, tax-free holidays and infrastructural provision. These foreign branch plants produce goods for the overseas market, manufacturing such items as ice hockey equipment and garments, for instance.

The policy was pioneered in Puerto Rico, with its 'Operation Bootstrap', and was followed by similar moves elsewhere in the Caribbean, as for example with Barbados' 'Operation Beehive' launched in 1969. This provided for tax-free holidays and the duty free importation of all materials for overseas companies establishing branches on one of the nine Industrial Development Corporation built industrial estates, a case which is fully detailed in the next section. Although in their infancy, both St Lucia and St Vincent and the Grenadines have followed similar paths, focusing respectively on their Hewanorra and Camden Park industrial estates. Harrison (1984) has noted that by the early-1980s, manufacturing contributed 12 per cent to the GDP of St Vincent, this having doubled in a mere six years. In the Dominican Republic there are now eighteen operational industrial Free Zones, with four under construction, and eight additional projects under active development. These zones provide jobs for over 120,000 people, representing nearly 28 per cent of the total labour force (Nanita-Kennett, 1998).

However, there are costs as well as gains involved in programmes which are based on foreign ownership and control. Operation Beehive, for instance, involved the establishment of some 173 transnational corporation branch plants in Barbados, and Kowalewski (1982) argues that such companies promote the interests of elite groups, whether these are local or overseas, thereby diverting attention away from the needs of the poor. A major problem is that when the tax concessions run out, or profitability is otherwise reduced, multinationals are likely to move swiftly and without ceremony in closing branch plants, without having to give any thought to the social consequences of their actions. This is exacerbated by the size and financial might of many First World corporations, which have meant that they have been able to exact favourable terms in their negotiations with Third World governments.

Additionally, such developments by their very nature are based on the creation of low-paid jobs, frequently for women and on a part-time basis. They may well involve very poor working conditions, as noted by Kelly (1986) in relation to St Lucia (see also Barry, Wood and Preusch, 1984). Further, Nanton (1983) notes that two firms actually persuaded the Government of St Vincent and the Grenadines to break its own minimum wage legislation by demanding and obtaining a "training rate" some 30 per cent lower than the statutory minimum then in force. Kelly's paper on St Lucia also points to the fact that many of those working in factories do so in order to acquire skills which they hope will subsequently help them to emigrate to a metropolitan country. Many of the same arguments apply equally well to the new growth area of the economy – that of offshore data processing in locations like Barbados (Potter and Clayton, 1996). All of these developments may be seen as correlates of the global process of divergence in the realm of production.

The other growth sector of the economy has, of course, been tourism (Bryden, 1973; Conway, 1983) and governments throughout the Caribbean have offered fiscal and financial incentives to encourage hotel construction (see Potter, 1983). With respect to the potential development impact of tourism, it has been estimated that as much as 70 per cent of all tourist expenditure in the Caribbean region is repatriated, principally due to the very high levels of foreign ownership of hotels (Fraser, 1985). For example, in 1970, eleven of St Lucia's fifteen principal hotels were owned by North Atlantic transnational corporations. Similarly, it is recorded that while by 1974, tourism provided some 400 jobs in St Vincent, 60 per cent of the hotels on Bequia were US owned. In the same year in St Vincent, 65 per cent of employment and over 70 per cent of hotel income were generated by ten foreign-owned hotels. Similar statistics are recorded throughout the Caribbean, so that in Barbados in 1980, some 74 per cent of Class 1 hotels were owned by foreigners (Potter, 1983). Of course, major restrictions may be placed on the indigenous population if large tracts of land, or indeed entire islands are purchased by overseas interests. In the Grenadines, for example, both Mustique and Petit St Vincent were purchased by foreign companies, whilst Mayreau came to be owned by a local landlord. Some 17.5 per cent of Bequia is owned by one US company. In the case of Mustique, it is documented that the local population was not allowed to own property and their birth and burial rights were curtailed in case they should claim rights to the land (Nanton, 1983).

The main thrust of the account so far has been to demonstrate that despite efforts to engineer economic change by enclave industrialisation and the promotion of tourism, the economies of Caribbean territories have remained characterised by highly constrained and limited production possibilities. Indeed, in so far as developments in the tourist and manufacturing sectors have strong connotations of dependency and emulation through the demonstration effect, it can be argued that they have in all probability served to increase the import of commodities for consumption. At the same time, the production of primary staples for export markets has remained as the main goal of production. Such developments have also led to continued location within the existing urban-coastal belts of the territories concerned, as shown in chapter 2.

Globalised Industrialisation: The Case of Barbados

Barbados as a small island state (431 km^2) in the Eastern Caribbean, which was formerly a British colony, achieved independence within the Commonwealth in 1966. The population of the island was estimated at the end of 1992 as 259,300. It is now classified by the World Bank as a Middle-Income Developing Nation and in 1996, its Gross Domestic Product (GDP) per capita at factor cost was US\$ 9,800. In the 1950s, Barbados introduced fiscal incentive legislation for potential overseas investors, beginning in 1951 with the Pioneer Industries (Encouragement) Act. This was followed in 1958 by the Pioneer Industries Act, which granted the manufacturers of designated products a seven year tax benefit (Potter, 1981). In 1963 these two original Acts were replaced by two more which extended the tax free holiday to ten years. In 1969 what had been created in 1955 as the Barbados Development Board became the Barbados Industrial Development Corporation (IDC), which was responsible for the development of manufacturing industry. The Revised International Business Companies Act of 1991 offered a further range of fiscal incentives to potential investors. Companies with no more than a 10 per cent local interest and which cater exclusively for markets outside the CARICOM region pay corporation tax at 1 to 2.5 per cent and can import free of duty all equipment, machinery and raw materials necessary to the functioning of the business.

In the early-1980s, 173 enterprises had been established, most of them located on the ten industrial parks established by the IDC. These are shown in Figure 3.1. By 1993, this total had increased to 217 (Table 3.1). Of these, 68 were either owned by foreign enterprises or were joint ventures between foreign and local firms. As well as the establishment of branch plants, TNCs also sub-contract work to local firms. For example, two local firms are involved in bottling carbonated drinks for Coca-Cola and Pepsi. The main foreign investor in export-oriented enterprises in Barbados is the United States. By 1992, manufacturing contributed Bds\$ 203.2 millions to GDP, and manufacturing employed 11,551 people, 7,265 of these being employed by foreign-owned firms assisted by the IDC. In 1991, the Census gave the economically active population of Barbados as 187,791.

Many of the industrial parks established by IDC from 1959-60 have seen overall expansion since the early-1980s, especially those at Grazettes,

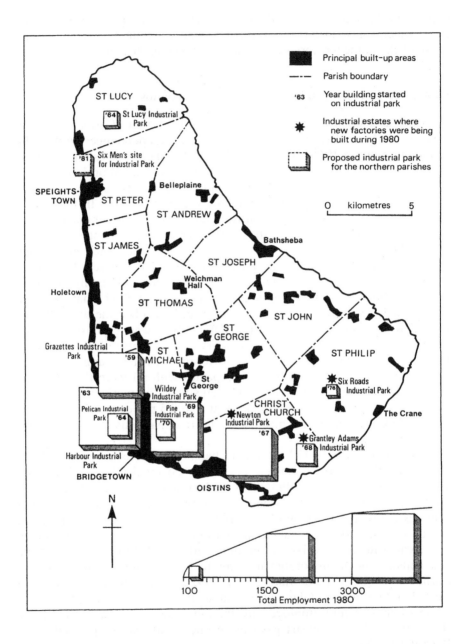

Figure 3.1   The ten industrial parks established in Barbados by the Industrial Development Corporation

Table 3.1 Industrial enterprises by location and employment on the industrial estates of Barbados, 1993

Industrial Park and Parish	Total number of enterprises	Total employment
Grantley Adams, Christ Church	12	470
Grazettes, St Michael	20	561
Harbour, St Michael	21	2,934
Newton, Christ Church	23	1,155
Pelican, St Michael	28	158
Pine, St Michael	15	127
Six Road, St Philip	6	118
Spring Gardens, St Michael	7	113
St Lucy, St Lucy	1	193
Wildey, Christ Church	20	700
non-IDC tenants	15	312
non-park	49	1,324
Total	217	7,265

Source: Clayton and Potter (1996)

Pelican, Grantley Adams and Pine (see Figure 3.1), whilst a new facility has been established at Spring Gardens to the immediate north of Bridgetown along the highway. The leading forms of activity are the production of food, beverages and tobacco (30), textiles and apparel (37), wood products (24), fabricated metal (24), electronics (12), handicrafts (21), and more recently, data processing (12). The parks show considerable specialisation, as indicated by the data for 1992 shown in Table 3.2. For instance, 43.5 per cent of the enterprises on the Newton estate are electronics firms, 75 per cent of those at Pelican are local handicraft businesses. The Harbour Industrial Estate is becoming the main location for data processing companies. At present, 25 per cent of firms here are in the computer data entry business. Recent data show that the electronics industry has declined in both output and employment, the latter by as much as 9.8 per cent between September 1991 and September 1992. At the same time, data processing has emerged as the new growth area. Whilst in 1992, data processing firms accounted for only 5.5 per cent of all companies assisted by IDC, the sector employed 23.8 per cent of the workforce.

This development started in 1983, when American Airlines moved its data entry operations from Oklahoma to Barbados, estimating that this would cut its costs by as much as 50 per cent. Caribbean Data Services, as the operation is now known, currently employs 1,045 workers, 92 per cent of whom are female, with an average age of 25 years. Salaries are said to be less than one-fifth of their American equivalents. Caribbean Data Services is a wholly-owned subsidiary of AMR Corporation, which is the parent company of American Airlines. By 1993, there were twelve data firms in Barbados, six owned by American or Canadian companies. Together they provide 1,800 jobs (58 per cent accounted for by Caribbean Data Services), processing credits for novels, insurance and health claims, telephone directories and airline tickets.

In an attempt to identify the motives behind companies locating in Barbados, managers at seventeen of the 68 foreign-owned firms were interviewed (see Clayton and Potter, 1996). Eight were American owned, five Canadian, one German, one European, and two Canadian-Barbadian joint ventures. These represented the electronic equipment, data processing, textiles, and several other sectors, as shown in Table 3.3. From the data collected it is clear that three main reasons were cited for locating in Barbados: its infrastructure (road, air, transport, telecommunications and electricity supply systems) (76.5 per cent), the level of literacy (47.1 per cent), and the cost of labour (41.2 per cent). The first two factors are not unexpected, but neither is the third in general comparative terms. However, labour cost is not such an important locational factor because of

Table 3.2 Number of enterprises by sector for the industrial estates of Barbados

	Grantley Adams	Grazettes	Harbour	Newton	Pelican	Pine	Six Road	Spring Gardens	St Lucy	Wildey
				Industrial estates						
Food, beverages and tobacco	3	3	3	4	0	2	1	0	0	7
Textiles, apparel and leather	1	6	6	1	2	8	1	0	1	5
Wooden products	2	4	0	2	0	0	2	5	0	0
Paper products and printing	0	2	0	1	2	2	0	0	0	5
Chemicals and toiletries	2	0	2	0	0	0	0	1	0	0
Plastic products	0	0	1	0	1	1	0	0	0	0
Non-metallic mineral products	0	0	0	0	0	1	0	0	0	2
Fabricated metal products	3	3	2	3	0	0	1	1	0	0
Precision instruments	0	0	0	1	0	0	0	0	0	0
Electronic equipment	0	0	0	10	1	1	0	0	0	0
Handicraft	0	0	0	0	21	0	0	0	0	0
Data processing	1	1	5	0	0	0	0	0	0	1
Other manufacturing	0	1	2	1	1	0	1	0	0	0
Total	12	20	21	23	28	15	6	7	1	20

Source: Barbados Industrial Development Corporation and *BIDC Directory*

Table 3.3   Principal reasons cited by a sample of foreign-owned companies making the decision to locate on the industrial estates of Barbados

Type of firm	Labour costs	Literacy levels	Proximity to US market	Infrastructure	English-speaking	Large labour reserve	Political stability	Other	Do not know
Electronic equipment			✓	✓					
Electronic equipment	✓	✓		✓					
Electronic equipment	✓			✓			✓		
Electronic equipment		✓		✓			✓		
Electronic equipment		✓		✓	✓		✓		
Data processing		✓		✓				✓	
Data processing	✓	✓		✓	✓	✓	✓		
Data processing		✓		✓	✓	✓	✓	✓	
Data processing									✓
Data processing		✓		✓	✓				
Data processing	✓							✓	
Other manufacturing	✓		✓	✓			✓		
Other manufacturing	✓			✓					
Textiles, apparel, leather	✓			✓				✓	
Graphites		✓	✓					✓	
Precision instruments	✓		✓	✓					
Paper products and printing									✓
Total	7	8	4	13	4	2	6	5	2

Source: Clayton and Potter (1996)

the relatively high wage levels of Barbados, so that companies locating solely on the basis of labour cost can find much cheaper locations elsewhere.   For example, a German-owned company which makes hairsprings for instrumentation, also has a branch plant in Portugal.   The firm reported that wage levels are 8.7 per cent lower in Portugal than in Barbados.   However, the manager stated that "although the wages in Barbados are higher, the quality of the finished good is much better here".

Looking at the results by sector, several trends are discernible.   All of the electronics firms cite infrastructure as the principal reason for locating in Barbados.   The data processing companies, with just two exceptions, state the importance of high literacy levels.   With the exception of one electronics firm, they are also the only sector to have taken the decision to locate in Barbados on account of the national language being English. Both of these attractions are easy to understand in the context of the training and accuracy requirements of data entry operations.   Interestingly, no company cited the fiscal incentives offered as a reason for locating in Barbados, no doubt reflecting the uniform nature of such incentives across Developing Countries.   The 'other category' included in Table 3.3 generally refers to much more personal reasons for locating in Barbados, including being on the route network of American Airlines, the influence of attending a trade convention hosted by the Prime Minister of Barbados, the role of IDC, and the opening of the 'ABC' (Adams-Barrow-Cummings) Highway, linking the airport to the west coast.

In conclusion, notwithstanding the notable successes in attracting industry over the past twenty years, it seems clear that, due to the realities of global divergence, a small country such as Barbados has to be highly cognisant of its precise position within the globalised system of manufacturing production, and that it faces strong competition within the global market place.   Such competition is bringing about a highly differentiated set of competing nations.

Patterns of Global Convergence and Contemporary Caribbean Urbanisation

The other principal aspect of the argument presented by Armstrong and McGee (1985) involves stressing the salience of what may be regarded as the increasing *convergence* or similarity which is occurring between countries with regard to patterns of consumption, as mediated through the dominant capitalist mode of production.   In this connection, the importance of the role of the mass media, especially television and advertising, in spreading patterns and aspirations of consumption that are typical of western countries, cannot be overstressed. This occurs both as a result of advertising campaigns, and through the lifestyles and aspirations which are portrayed on a daily basis. It is, of course, axiomatic that cities play a fundamental role in the diffusion of new techniques, products and fashions, both internationally and within nations, but in a manner which is far more complex and subtle than conceptualised until quite recently (see Potter, 1990).

Interestingly, this accords well with the arguments which have raged since the late-1960s, concerning the manner in which innovations are diffused within national and regional urban systems. During the era when modernisation theory dominated developmental thinking, authors such as Hudson (1969), Pederson (1970) and Berry (1972) argued that new techniques and processes diffuse inexorably down the urban size hierarchy. In such a formulation, development is merely a question of the time lag which is involved in the spread of growth-inducing changes. However, as shown so clearly by Pred (1977) in relation to the United States, in the contemporary global economic setting, innovations are increasingly being bounced around between the large urban areas which comprise the upper levels of the global urban system.

Further, it is notable that the illustrations of the process of hierarchical diffusion provided by Berry, all related to domestic-oriented innovations such as television stations, that is traits which are closely related to patterns of consumption. In contrast, it can be suggested that it is entrepreneurial or production-oriented innovations that are held within the upper levels of the urban system. The role of multinational companies in this process and the operation of an urban size ratchet effect are of considerable importance in this respect. Clearly, this conforms with the argument that increasing convergence is occurring with respect to patterns of global consumption.

It can be argued that urban areas are involved in the concentration of capital, ownership and production on the one hand, and the simultaneous diffusion of patterns and processes of consumption on the other. The common or garden expression of these latter processes is witnessed in the similarity which is observed between large stores and high-rise office, apartment and hotel complexes in the towns and cities of both First and Third World countries. The same trend is witnessed in the increasing dominance which is shown by roads and cars in relation to evolving urban forms. It is yet again manifest in the development and growth of elite residential zones. The role of the professions, for example, government economists and planners in this process, especially those who have been trained abroad, must not be overlooked (Potter, 1985). Zetter (1981) has, for example, argued that the imposition of the norms of planners who have been trained overseas on the indigenous masses is likely to give rise to culture clashes of the highest order. Although it can be argued that it is the well-to-do groups that are involved most widely in these processes, their real significance results from the fact that they affect many aspects of the daily lives of the poor. Prime instances are provided by the penetration of Third World markets by tobacco companies, as well as those selling soft drinks, fast foods and formula milk for babies. In respect of the latter, for example, in a study of two hundred mothers in St Vincent, Greiner (1977) demonstrated that early weaning and the supplementary use of infant food among poor mothers were the two factors most connected with malnutrition. This prompted the author to suggest that on the basis of the evidence presented, it would be justifiable to ban all forms of infant food advertising in St Vincent and the Grenadines.

Leaving aside the all important issues of political, military and economic hegemony, it is these traits which are leading to an increasing 'Americanisation' of many Third World regions. It also explains why, superficially at least, towns and cities in, say the Americas and indeed elsewhere, increasingly look the same, in certain respects at least. Potter and Dann (1987) for instance, have argued that Barbados, long described as 'Little England', can now be properly understood as 'Little America'. At the level of the individual urban area, Clarke (1989) has described Kingston, Jamaica as a western implant. One trend that was observed early on in the case of Latin American cities was the marked tendency for the elite residential area to migrate out from the traditional pre-industrial core in a sectoral manner. The existence of such 'Hoytian' sectors has been well documented in the case of Bogota, Quito, Lima, Santiago and Caracas by Amato (1970a, 1970b), along with others (Morris, 1976; Lowder, 1986). The trend is so pronounced that it has been mapped into a generic model of Latin American city structure by Griffin and Ford (1980). They give prominence to the occurrence of a high-grade commercial linear spine which extends out from the CBD. This is the focal point for the location of not just the best shops and most prestigious office buildings, but also of museums, theatres, zoos and parks. Around this zone are to be found the homes of the elite. The remaining areas of the city are deployed on a concentric basis, with there being a general diminution in social standing with distance travelled from the urban core. This modern equivalent of the patterning found within the pre-industrial city reflects the process of in situ upgrading which has occurred within the generally older squatter settlements and shanty towns which are located close to the urban core. More recent and less consolidated informal settlements are to be found on the periphery (see Potter and Lloyd-Evans (1998) for further details).

There is some evidence that within Caribbean urban areas, affluent groups are moving out of the city toward the suburbs in a sectoral manner as in many North American and South American cities. In a review of urban development in Nassau, Bahamas, Boswell and Briggs (1989), although arguing that the land use pattern does not fit neatly any of the standard models, noted the existence of commercial strip developments which extend both eastwards and westwards from the CBD. The western strip development is comprised of mainly hotels and restaurants, whilst the western residential sector of the city is characterised by a mixture of commercial and high-income residential uses.   It is noted that the businesses in this area are primarily aimed at the tourist. The emergence of a new middle-class and an American-style flight to the suburbs in the 1960s and 1970s has been documented in the case of San Juan, Puerto Rico (Albuquerque and McElroy, 1989), and the intensity of suburbanisation-metropolitanisation has decreased sharply over the past ten years.

Even in a medium-sized urban area such as Port-of-Spain, Trinidad, the structure and development of the city has been described as a history of sectoral expansion by Conway (1981) (see also, chapter 2 of this volume). In particular, a high-density, low-income residential district has developed, radiating from the central city to the eastern periphery. It includes areas such as Belmont, Success Village, Eastern Quarry/Prizgar Lands, Troumacaque and Chinapoo (see Figure 3.2). Within this area, three zones can be recognised: a central tenement area, a ring of early established spontaneous settlements, and a new periphery of uncontrolled settlements (for example, Morvant).   Meanwhile, wealthy urbanites have relocated farther away from the commercial core by leap-frogging the eastern city, or else have developed suburban residential areas in the northern and northwestern valleys, such as Diego Martin (see Conway, 1981).   The basically sectoral social stratification of contemporary Port-of-Spain is witnessed in the recent analyses of the 1980 Census data. Figures 3.3 and 3.4 show indices of housing amenity derived from a multivariate analysis of a range of housing variables (Potter and O'Flaherty, 1995).

The derivation of high-status sectoral suburban wedges in the post-war period is also typical of many smaller Caribbean towns and cities.   In Bridgetown, Barbados, for example, distinct northern and southern high-status private residential zones have developed (Potter, 1989a, 1992a), as shown in Figure 3.5. Even in Kingstown, the capital of St Vincent and the Grenadines, there has been strong suburban residential development to the southeast of the capital between 1980 and 1991 (Potter, 1992b). In the case of some of the smaller islands, the gateway origins of most Caribbean urban centres has affected morphology more directly, for the chief consideration was their function as sheltered maritime ports which could be defended with ease. Hence, as noted in Chapter 1, the typical urban site was on the shore of an embayment, protected on the landward side by

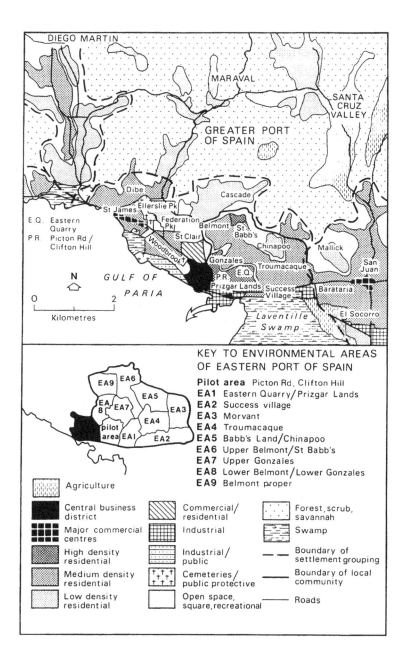

Figure 3.2   The structure of Port-of-Spain, Trinidad

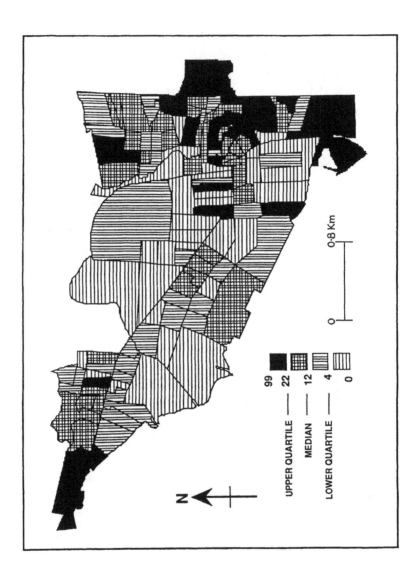

Figure 3.3   Contemporary housing conditions in Port-of-Spain: the proportion of houses constructed of wood

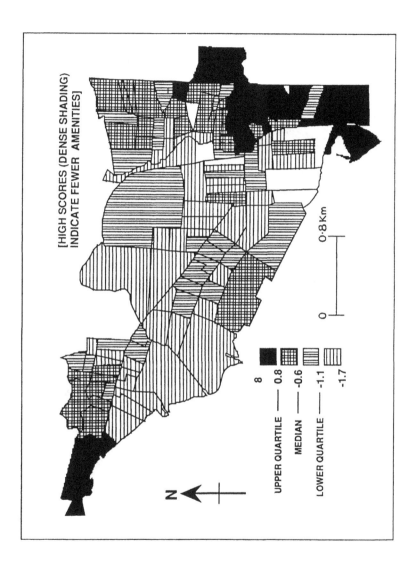

Figure 3.4   Contemporary housing conditions in Port-of-Spain: an overall index of amenity

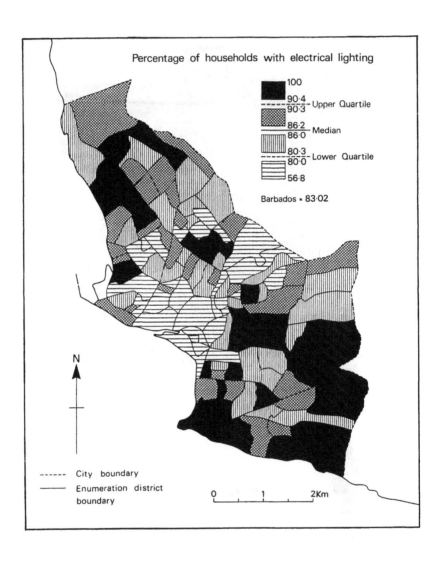

Figure 3.5   Housing conditions in Bridgetown, Barbados

commanding hills on which forts overlooking the sea approaches could be built (West and Augelli, 1976). This has given rise to rather cramped and huddled urban areas, such as Castries (St Lucia), Kingstown (St Vincent and the Grenadines), St George's (Grenada) and Roseau (Dominica). In these smaller towns, the high-status areas frequently developed around these relatively cool and commanding heights.

The Caribbean has participated in the rapid post-war growth that has occurred in international tourism, and the region now accounts for approximately 3 per cent of the global market. As suggested in the section on patterns of production, much of recent Caribbean urbanisation has been strongly predicated on tourism. As well as the foreign ownership of hotels, evidence points to the fact that the growth of tourism often exacerbates dependence on imports, particularly of foods. Equally, research suggests that the tourist sector dampens the availability of labour for agriculture. The competitive bidding up of land prices in the tourist-oriented small islands of the eastern Caribbean is also well documented.

Another consequence of the industry is the so called international demonstration effect, which involves the rapid assimilation of affluent North American tastes and consumption patterns (see McElroy and Alburquerque, 1986). One of the chief issues raised in this connection is the likelihood that faster rates of change will characterise consumer preferences than productivity. Hence, many fear that given the limited insular resources available to such small islands, a totally inappropriate level of adoption of imitative metropolitan lifestyles may be occurring. This may lead to the pernicious adulation of all things foreign and the fostering of continued psychological dependency (see Beckford, 1972).

Such trends have also affected the built environment, for tourism, whether associated with modernist or postmodern architectural designs, has focused yet more attention on the urban-coastal-littoral zone. Thus, Albuquerque and McElroy (1989) refer to the 'Miamization' of San Juan, Puerto Rico. In Barbados, there are now precious few gaps between tourist facilities on the seaward side of the west coast highway which runs from Bridgetown to Speightstown. Even in the case of Cuba, there are current fears about the level of prostitution and dollar trafficking which have been attendant upon the recent resurgence in tourism. Others, including local architects and planners have bemoaned the tendency toward uniform modernist architecture that the development of the industry in the past has involved, and the case for redevelopment along more 'traditional' post-modernist lines has been set forth by Segre (1991). Many of these themes are further developed in Chapter 4.

Another distinct expression of the tourist demonstration effect is represented by changes in indigenous dietary preferences. MacLeod and

McGee (1990) have recently referred to the emergence of the 'industrial palate', whereby a larger proportion of food is consumed by non-producers (see also Drakakis-Smith, 1990). There is more to this than the spread of fast food stores such as Kentucky Fried Chicken, MacDonalds, and the ubiquitous impact of Coca-Cola and Levi jeans, although of course, these trends cannot be dismissed. Such changes have also been associated with the rise of supermarketing as a form of mass merchandising. In Barbados, for example, a staggering 97 per cent of total national retail floorspace is located within the western and southern linear urban corridor, while this area accounts for 77.93 per cent of total population. The distribution of retailing facilities in Barbados is shown in Figure 3.6. But the most salient fact is that the middle ranks of the national retailing hierarchy are entirely accounted for by nine new centres which are to be found located in, and around, suburban Bridgetown (Figure 3.6). All of these have been developed since 1965, and they are based on relatively large-scale supermarkets. They primarily cater for expatriates, tourists and the indigenous upper-middle and upper classes (see chapter 4). They sell a very high proportion of imported foods and wines. This array of centres was added to in December 1989, when Plantations Limited opened a superstore in Six Cross Roads at the southeastern end of the urban coastal belt. This consisted of some 46,292 square feet, including lumber sales. However, this closed down just a few years later. Noticeably, outside the historic linear urban belt, there are no significant retail concentrations, and there are precious few modern facilities in any of the district centres. These areas are principally served by rural rum shops, plus a scattering of small mini-marts.

Whilst convergence on western patterns of consumption is so clearly becoming the societal norm for Caribbean territories as a whole, as argued previously, it must be fully appreciated that the 'participation' of different social groups in these developments is highly uneven. This is shown by the case mentioned above, whereby the poor are disproportionately dependent upon local retail facilities, where prices are likely to be appreciably higher. In this sense, the continued existence of 'traditional patterns' of consumption must be recognised alongside modern forms. Thus, whilst in many areas of consumption the *dissolution* of traditional forms has occurred, this has not been the case in all instances, especially for the poor. In these circumstances, the *conservation* of pre-existing forms is of great importance.

The same is also true with regard to the consumption and production of housing. In recent research, the present author has examined in detail, housing quality in the eastern Caribbean region (Potter, 1991a, 1991b, 1991c, 1994; Potter and Conway, 1997). None of the Windward Islands

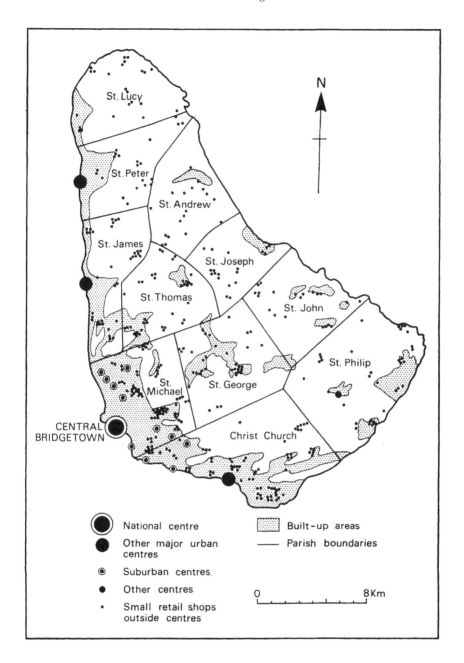

Figure 3.6   Retailing facilities in Barbados

possess a national housing plan, nor do they support effective Ministries of Housing. Rather, in each, a technically-oriented national housing authority is vested with the task of improving housing on the basis of profitability and cost recovery. The result has been that virtually no houses have been built, and very few have been upgraded. The failure to produce explicit housing policy statements can, of course, be seen as an implicit state policy of letting the poor fend for themselves. As implied above, this can be interpreted in the light of ideas emanating from articulation theory, concerning modes of production under capitalist development (Armstrong and McGee, 1985; Burgess, 1990; Potter, 1992b). This argument is further developed and elaborated in chapter 6.

The stance of governments may be seen as one of primarily seeking to conserve the 'precapitalist' or 'informal' housing system in its most basic of forms. Both the fostering of strict self-help, involving minimalist systems of state assistance, and brief flirtations with complete state housing systems (consisting of western-style blocks of flats or row housing) which are then abandoned, can be seen as leading to the ultimate conservation of the folk/vernacular architectural form at its most basic level (see chapter 6). However, the intermediate position of improving such housing by means of appropriately financed state aided self-help housing has been almost totally ignored. This might involve the provision of small, low-interest loans and technical assistance on site. It would almost certainly entail producing wet core units associated with wooden houses built along traditional lines (Potter, 1991c, 1992a, 1995). It would witness the enhancement of the local vernacular architectural system, not its maintenance at a basic and exploitative level (Potter and Conway, 1997).

Concluding Comments

This account leads us to refer back to the model of the principal development issues which the small islands of the Caribbean face which was presented in chapter 1 (see Figure 1.4). The exogenous realities for Caribbean territories consist, on the one hand, of the extraction of social surplus product, both by means of the export of agricultural staples and raw materials at minimum market prices, and expenditure, investments and transfers abroad by elite groups. In the other direction, flows of imported foods, manufactures, tourists, enclave industries, offshore data processing, aid, new technology and consumerism are characteristic. The Caribbean urban nexus is influenced directly by the processes of globalisation, especially those of convergence and divergence, but also of postmodernity. These forces must be added to the longstanding ones of dependency set within a contemporary framework of structural adjustment, deregulation and cost recovery (Figure 3.7).

The geographical pull of accessible and well-developed areas with good infrastructural facilities for industry and of safe and scenic beaches with regard to tourism have served to confine recent developments to those very same leeward coastal tracts that centuries earlier had first attracted mercantile capital. The continued channelling of new developments into these 'relative economic oases' can be seen as having left counterbalancing 'economic deserts' (see Figures 1.4 and 3.7). But the salient point is that the socio-economic contrasts between rural, urban and suburban areas are real and affect the daily lives of citizens, but are predicated on strong urban-rural flows which involve unequal exchange. As noted in chatper 1, this takes the form of low procurement prices for agricultural products, and low wages in the tourist and domestic sectors of the economy. The disadvantageous terms of trade that affect Third World countries at the global scale are thereby matched by harsh urban-rural terms within countries. This is reflected in long journeys to shop, to places of education, health centres, government offices and all manner of other facilities (for empirical details see, for instance, Potter, 1989c). But in the case of island micro-states territorial size and the historical legacy of concentrated infrastructural provision may well be cited as the justification for maintaining, or even enhancing a highly polarised pattern of spatial development. The model is obviously a highly simplified one, but should act as a starting point for the consideration of processes of urban and regional change in the Caribbean region.

It should, of course, be noted that states which have espoused an overtly socialist path to development in the Caribbean have made the effort to reduce the apparent schism between the urban and rural components of the national territory one of their prime concerns (see Potter, 1993a, 1995). This is nowhere better exemplified than in the case of post-revolutionary Cuba, where the overall remit of the state has been to urbanise the countryside and ruralise the towns (Gugler, 1980; Susman, 1987). The main policy thrusts of the Cuban state have been directed towards increasing production rather than consumption, and to effecting a functional integration between town and country. Most salient has been the remarkable efforts made to redistribute social surplus product to rural areas by the provision of health and education facilities. Essentially similar efforts were made by the People's Revolutionary Government (PRG) under Maurice Bishop in Grenada. On the eve of the revolution of 13 March 1979, Grenada suffered from a chronic trade deficit, strong reliance on grants in aid and remittances, heavy dependence on food imports, high unemployment and extensive tracts of idle land (see Brierley, 1985). The

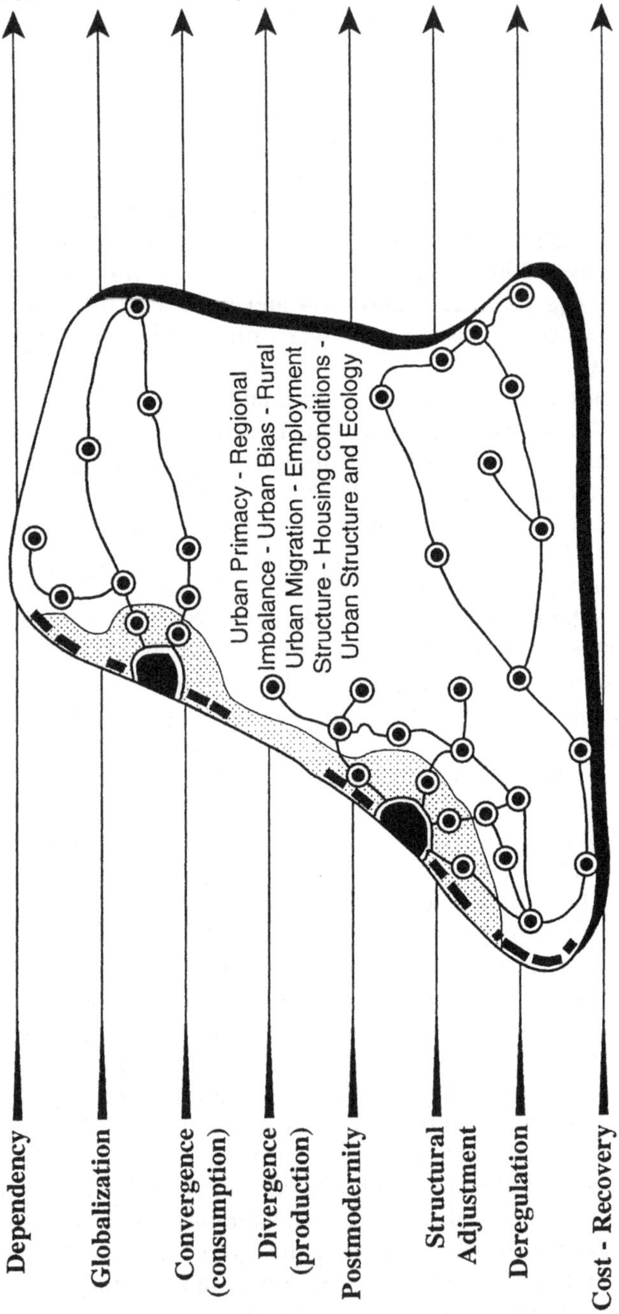

Dependency
Globalization
Convergence (consumption)
Divergence (production)
Postmodernity
Structural Adjustment
Deregulation
Cost - Recovery
Urban Primacy - Regional Imbalance - Urban Bias - Rural Urban Migration - Employment Structure - Housing conditions - Urban Structure and Ecology

Figure 3.7 The current array of factors affecting Caribbean urbanisation

PRG immediately froze food prices and declared its intention to depart from the traditional role of Grenada as the exporter of cheap agricultural products. With respect to agriculture, the aim was to achieve greater self-sufficiency. In this connection, a major effort was made to 'marry idle hands with idle lands'. With the opening of the new Point Salines international airport, the PRG also declared its intention to develop what it referred to as the 'New Tourism' - the term given to sociologically more relevant forms of vacation activity, stressing the culture of the country, along with indigenous cuisine, handicrafts and furniture-making. It should be stressed that although seen a 'socialist' revolution, 80 per cent of the Grenadian economy remained in the hands of the private sector. Even the World Bank commented favourably on the overall level of growth recorded under the PRG between 1979 and 1983.

In conclusion, it must be emphasised that given the operation of the seemingly contradictory processes of conservation and dissolution, it would clearly be naïve and downright erroneous to regard Caribbean towns and cities as evolving replicas of those found in North America, or indeed anywhere else. The forces of conservation and dissolution are inexorably leading to greater relative differentiation within society and in the built form of Caribbean urban settlements. First and foremost, the nature of urbanisation in such peripheral capitalist societies depends on the precise balance that is struck between global convergence and global divergence on the one hand, and conservation and dissolution on the other.

4 Tourism, Post-Modernity and the Caribbean Urban Imperative

ROBERT B. POTTER and GRAHAM M.S. DANN

Introduction

The previous chapters have illustrated, in broad outline terms, the salience of the concepts of modernity and post-modernity to any consideration of urban change, globalisation and development in the contemporary Caribbean. This proved to be particularly the case when considering the updated and revised version of the plantopolis model reviewed in chapter 2. In addition, concepts of modernity/post-modernity also entered into the discussion of urban structures presented in chapter 3. But post-modernity is particularly related to the promotion and development of tourism in the Caribbean, and as already argued, tourism is a strong impetus to present-day urbanisation in the Caribbean. This theme is fully explored in the present chapter. At a number of junctures, following the arguments presented in chapters 2 and 3 of this volume, the implications of such themes for urban change and urban structuring in the wider Caribbean region are explored. In chapter 5, the debate is broadened out to consider the wider environment.

Tourism and Post-Modernity: An Overview

Following the implosion of the post-modern perspective on to the social sciences in the 1980s (Denzin, 1986; Featherstone, 1991; Harvey, 1989; Jameson, 1985; Lash, 1991), it was only to be expected that such a theoretical framework would soon find its way into the analysis of contemporary tourism. Beginning with MacCannell's (1976) seminal treatment of the tourist as an alienated individual seeking authenticity in the culture of the "Other", a sort of present-day pilgrim journeying towards self-identity, only to have this "sacred quest" thwarted by the machinations of the tourism industry (Boorstin, 1987), there have been several works which, via critique and challenge, have attempted to develop and refine some of these ideas (see, for example, Cohen, 1988, 1993; Crick, 1985, 1989a; Graburn, 1992; MacCannell, 1992; Selwyn, 1992a,

1992b; Urry, 1990, 1991).

Of course, tourism itself as emblematic of hyper-reality and fantasy (Eco, 1986), as the vehicle *par excellence* of created imagery, commodification (Cohen, 1988), consumption of myths (Dufour, 1978; Selwyn, 1992b), de-differentiation of culture and the nostalgic re-creation of heritage (Dann, 1994; Frow, 1991; Urry, 1990), becomes in many ways the perfect symbol of the post-modern ethos. Tourism, through its multi-national hotel chains and central reservation systems (Poon, 1993), more than any other mass phenomenon oversteps boundaries, internationalises entire sovereign states (Lanfant, 1980), produces its own network of *ad hoc* centres, and creates fuzziness between the 'consumers' and the 'consumed'. Tourism uproots place by thematising the natural (Cohen, 1993), and playfully reconstructs time by selectively staging history and events (Cohen, 1985, 1986; Moore, 1980; Allcock, 1992; Urry, 1990). As Selwyn (1992a: 129) so aptly summarises the situation:

> Sites are no longer signifiers which, linked together, form coherent structures within which individual tourists find historical and biographical meaning, but instead they are centres of physical and emotional sensation from which temporal and spatial continuities have been abolished.

Caribbean micro-states are no exception to this current worldwide trend. Advertised as "nature" islands (Widfelt, 1993), and as tropical havens for rest and relaxation, what are Third World countries or nations of the south become defined and transformed into ludic centres of hedonism (Lett, 1983) for the conspicuously consuming golden hordes of escapist playboys and playgirls deplaning from the frozen north (Turner and Ash, 1975; English, 1986). But in truth these "nature islands" are frequently dominated by strongly urban concentrated hotel and tourist facilities.

Yet curiously, few writers have attempted to analyse the region's number one industry in post-modern terms. True, there is Lett's (1983) article on charter yacht tourism, and Fussell's (1979: 31) allusion to Jonestown (Guyana) and his discussion of this site of mass religious suicide as a possible attraction in "an advanced phase in the age of tourism" as well as his reference to cruiseships as "movable pseudo places making an endless transit between larger fixed pseudo places" (Fussell, 1979: 35). Thurot (1989: 33-35) speaks of Jamaica as a destination of touristic consumption temporarily inhabited by such globe-trotters as James Bond (cf. Cazes, 1976: 9), who are accustomed to palatial, frequently urban, surroundings while imbibing magnum after magnum of champagne and consuming vast quantities of smoked salmon. Gradually, through a system of verisimilitude, such Caribbean locales can be fabricated into a pastiche of Polynesian architecture, erotic voodoo and creole cuisine, which becomes defined as more real than reality, in just the same way as Puerto Rico, promoted and seemingly created by Bacardi, is transformed into a catwalk for models displaying the latest in *Vogue* fashions (Thurot, 1989:44).

Crick (1989a: 333), who argues that "tourism is a way of providing a simulacrum of the world", as instances of the identified image-advertising-consumption pattern of post-modernity, supplies examples from the Bahamas and their portrayal as the playground of the western world, South America and its description as an enchanted forest where Walt Disney's "Bambi" lived, and Greek fishing villages of the Caribbean incorporating the best of the Mediterranean of Mexico's Pacific coast (Crick, 1989a: 329). This sense of displacement (Relph, 1983) is experienced by Boorstin (1987: 98-99) in the Caribe Hilton of San Juan, incidentally the first establishment of the Conrad empire. According to Boorstin, the hotel as a "little America" could have been constructed anywhere. It is a model of "American modernity and antisepsis". The interior design could have been taken from an American airline. Even "the measured admixture of carefully filtered local atmosphere proves that you are still in the USA". Relatedly, Britton (1980: 47) quotes a Hilton official as saying: "without the large hotels, most of the Caribbean islands would dry up and blow away". The urban base of such contrived-virtual realities is noticeable.

English (1986: 48), who maintains that "North-South tourism injects the behaviour of a wasteful society in the midst of a society of want", refers to the subsequent "coca-colaization" of indigenous life, under which Third World natives strive to adopt the obvious marks of affluence shown by their visitors. These forces relate directly to global convergence as discussed in the previous chapter. He also alludes to the "smile campaigns" conducted in Jamaica, the Bahamas and Barbados, as forming part of the artificial welcome associated with such behaviour (English, 1986: 72).

Britton (1980: 45), who focuses on the escapist motivation of the post-modern tourist, reckons that it leads to over-indulgence in food, alcohol and sex and the chance to do things which are sanctioned at home. Thus, Jamaica becomes equated with ganja, and Caribbean culture in general with the steel drum (Britton, 1980: 46). No small wonder, he argues, that there is a concomitant disregard for place and people and that the latter turn to inappropriate behaviour and alien consumption models - having in the words of one West Indian Prime Minister "cadillac tastes in bicycle societies" (Britton, 1980: 45). Kincaid (1988: 15-16 cf. Errington and Gewertz, 1989: 52), too, refers to the post-modern condition of the tourist, though preferring to accentuate the pre-trip situation of loneliness, lack of love and displacedness. These are the feelings, she believes, which lead people to visit places like Antigua, where supposedly and imaginatively there is harmony and union with nature rather than *de facto* anger and oppression.

Many of these disjointed themes and threads, which emerge from one partial application of post-modern perspectives to an array of Caribbean territories, can perhaps be better understood in relation to one of its most mature tourist destinations - that of Barbados (Husbands, 1986). If McElroy and de Albuquerque (1989) are correct, this particular island, like Bermuda, the Bahamas and the US Virgin Islands, has evolved through various tourism "styles" to the point where it is typical of mass tourism locations facing the consequences of high impact/high density visitor intrusion. Now in the last stages of Butler's (1980) resort cycle, Barbados is confronted with the choice of rejuvenation or decline. For this reason, we have argued elsewhere (Dann and Potter, 1997) that any "attempts at revitalization should be understood as the efforts of a developing country to come to terms with 'Third Wave' societies (Toffler, 1980) from where the majority of its visitors originate". In the previous account we supplied various possibilities (for example, product diversification, niche marketing, encouragement of small guest houses) which have the potential of synchronising supply with changing tourist demand. In the present account we prefer to concentrate on host strategies which try to cater to the imputed post-modern outlook of the guest. Accompanying examples have been selected for their ability to highlight some of the previously elaborated themes. Together with commentary, they are briefly treated under five broad headings as follows: (i) Peddling Paradise; (ii) Commoditizing Culture; (iii) Hawking Heritage; (iv) Playful Placelessness; and (v) Blurring Boundaries.

Peddling Paradise

According to Cohen (1982), the contemporary touristic quest for paradise can be appreciated in relation to the traditional symbolic tension between the socio-moral order of the Centre and the undifferentiated state of nature of the "Other". Moreover, such imagery is also particularly appropriate to an understanding of the mythical qualities of small islands. Certainly the discourse of modern advertising has toned down the original religious meaning of paradise by converting it into secular equivalents, so that spiritual renewal is translated into rest and relaxation and self-fulfilment into the satisfaction of consumer desires. Yet publicity, nevertheless, promises that in seeking paradise - the far off place, the simple life, unspoilt nature and happy natives - all these attributes will be provided in paradise-like settings. In so doing, it responds to a romantic Fernweh/ Zeitweh in the tourist, the condition for transport out of place and time (Cohen, 1986; Wagner, 1977).

Cazes, (1987), whose analysis also focuses on tropical islands, indicates that the imagery and language of promotion typically concentrate on three processual themes: *appropriation* (virginity, emptiness, abundance), *insularisation* (refuge, haven oasis, fortress, cocoon, embryo) and *mythification* (the world caught between the real and unreal, emphasising bliss, freedom, absence of work, beautiful uninhibited women). He refers to a 1986 survey conducted by American Express in France, which showed that 60 per cent of the population had already been converted to the idea of an island as the most preferred type of tourist destination. Thus, he argues, even if the tourism industry cannot fully match such demand, it can at least provide island-like surrogates in the form of beach villages, clubs, marinas and leisure parks. In concentration, such post-modern developments will frequently bolster the existing urban littoral zone, thereby adding to tourist-based urbanisation.

In relation to Third World resorts, Cazes (1976) makes the point that the advertised images of tropical paradises are no more than the subjective projections of their potential clientele. Visiting such places thus becomes a tautological experience in which destinations are defined according to the values of their patrons. Tresse (1990), whose study of promoting French-speaking Africa could be equally applicable to the Caribbean, highlights a number of themes which are supposed to characterise the people of the former region - the *joie de vivre* (populations given over to partying), *simplicity* (in harmony with nature), *cultural richness* (abundance) and *community spirit*.

In a complementary fashion, Dufour (1978) stresses that the reason why brochurese is so successful is because it is couched in myth: the myth of the golden age, the horn of plenty, the fountain of youth, of Heliopolis, of Olympus, of Aphrodite and of Eros, indeed most of the Greek pantheon. He maintains that such heavenly discourse is particularly effective since it appeals to the "société de consommation ou post-industrielle" (Dufour, 1978: 12). There is a mystification and manipulation of myths by the mass media (Dufour, 1978: 32). It is a pre-logical language of archetypes - of Mother Earth, of valleys, caves, "la dolce vita" and gastronomy (Dufour, 1978: 23).

Finally, the theme of paradise is treated inversionistically by Gottlieb (1982) as the opportunity for members of the working class to be a "king" or "queen" for a day. This interpretation is particularly appropriate to the post-modern dissolution of "group" and "grid" (Lash, 1991: 26-30), and is further supported by post-structuralist readings of brochures which treat themes in a liminal fashion. Selwyn (1992a: 134-145), for instance, gives examples of "the coronation of the tourist" (the luxury hotel) and "the view from the throne" (looking out over the exotic kingdom).

In the Caribbean there is, of course, a real "Paradise Island" - a world of hotels and casinos just a dice throw from Nassau - (which for some may be closer to hell than to heaven). Thus, for those who wish a "perfect escape from the Caribbean" (Widfelt, 1993: 1), there is Pleasure Island. Originally Great Stirrup Cay, it was purchased from the Government of the Bahamas by Norwegian Cruise Lines, who subsequently changed its name and ensured that as few locals as possible were present to detract from the enjoyment of their passengers.

In Barbados, the paradise message from the then Board of Tourism (Travel and Leisure, 1990: 251) is rather different:

> Play the Bajan way...Only a select few know the special pleasures of Barbados and the charm of the Bajan people. These serene and gracious hosts will help you feel at home...in Paradise. See your travel agent to discover more about taking your vacation the Bajan way. FEEL THE BAJAN WAY. ONLY IN BARBADOS.

Here, instead of placing the accent on an uninhabited Eden far from the madding crowds, "the select few" are aware that pleasure is associated with the "charm" of the local people, who are affectionately designated as "Bajans". While it is true that the serenity and graciousness of the hosts conjures up Gottlieb's (1982) regal image of the sovereign with attendant servants, "feeling" the Bajan way suggests that maybe more is on offer, something closer to the "droit du Seigneur" which transcends the homespun sanctions of inter-racial encounters and tactile taboo. Ironically, the accompanying picture shows only two tourists and no Bajans. A young man is lying on a raft, looking beyond a girl's back to the open sea. Unlike their absent hosts, both visitors are white.

The kingdom theme is exploited once again in an on-site advertisement for one of Barbados' leading hotels - Marriott's Sam Lord's Castle. "Come play at our Castle" is the imperative (*Visitor*, 1993a: 7). The property is later described as having once hosted royalty, and hence is always seen as being "fit for royalty". Interestingly also, a recent refurbishment programme is said by the financial controller to have been in preparation for a "Kingdom for Hire" group visit in the Spring of 1994 which took over the whole establishment (*Barbados Advocate*, 1993d).

Wilson (1993: 12-13) refers to the advertisement of the Barbados Tourism Authority placed in the London *Sunday Times* of 31 October 1993, which read as follows:

The rich and famous have always looked with great affection on Barbados and have discovered a richer, kinder and more colourful world. Which is why - quite apart from an abundance of natural beauty - you'll find Barbados is ideally equipped for all the sports of kings and champions. Racing, yachting, golf, tennis, cricket, polo, surfing and scuba diving in the deep, myriad colour of our warm sea.

On this occasion royalty gives way to the simulacrum of the electronic media and the TV generated "rich and famous". There is now a play on the notion of richness - affluence in the visitor and a corresponding abundance in the sought after destination. True, the rich can engage in the sport of kings, but there are other sports as well. Those who can afford to play are welcome to pursue up-market activities.

The Eden theme, by contrast, is not applied to the whole island. Instead it is limited to one or two "natural" attractions. In this vein, Andromeda Gardens is referred to as "the loveliest, most interesting and varied tropical garden in the Caribbean" (*Barbados Holiday Map*, 1991). The associated timelessness of paradise is spoken of in terms of an odyssey, not as stepping back into the past, but rather as descending to the depths of the ocean, only to emerge as somehow born again or recreated (ie. analogous to the baptism imagery of Christianity). Thus in the Atlantis submarine pleasure craft one can:

> experience an unforgettable journey beneath the crystalline waters of Barbados...descend to a world of mysterious marine life, colourful reefs, a shipwreck and spectacular sponge gardens (*Visitor*, 1993a: 2).

This aspect of paradise is the primordial chaos, the rudimentary state of nature to which Cohen (1982) refers, the pursuit of Self through the Other, and the basis of religious symbolism (Eliade, 1968). "Untamed paradise" is also the theme of Barbados' Wildlife Reserve "situated in a lush mahogany forest, where uncaged animals are undisturbed in an idyllic environment" (Barbados Wildlife Reserve, nd).

Commoditizing Culture

From the multi-vocal theme of "paradise" one quickly reaches the post-modern equivalent in the consumer ethic and the related touristic imperatives of the "shopping paradise" (Cohen, 1982) and "bargain paradise". Indeed, Cazes (1976: 4, 41) refers to tourism an "un système de consommation fétichisée" in which the discourse of publicity material becomes a language of appropriation, where the "lieu de consommation" is soon reduced to "consommation de lieu". The link to the process of global convergence and the tourist demonstration effect discussed in chapter 3 is clear.

The heavy accent on purchasing is reflected in an underpinning marketing model, and the tourist brochures are replete with verbal and pictorial references to buying, shopping, bazaars, souks, and the ritual of bargaining (Cazes, 1976: 36). In considering the merchandising of tropical islands, Cazes (1987: 5-6) highlights a "discourse of consumption" which extends beyond the act of purchasing to the additional possibilities of renting and investing. The urban-coastal nexus is naturally prominent in this connection. Cohen (1982: 1), also in reference to Third World islands, alludes to the transformation of paradise "from utopian myth to consumer product", in which a commercial paradise springs up from within the (recreational) tourist's own cultural Centre, thereby responding to basic (late modern) needs. Contemporary mass tourism, he maintains, is in the business of selling destinations. The latter thus become paradise-like or contrived. What were once remote are now accessible. What was originally a religious quest is now converted into mass consumerism (Cohen, 1982: 8-9). The strong urban-basis of such imperatives can be appreciated by reference to Stage 3 of the plantopolis framework presented in chapter 2.

Crick (1989a: 308), who describes tourism as a "form of travel...made safe by commercialism", goes on to argue that "tourists do not go to Third World countries because the people are friendly; they go because a holiday there is cheap" (Crick, 1989a: 319). In this light, North-South tourism may be interpreted as a form of "conspicuous consumption in front of the deprived" (Crick, 1989a: 317). Rivers (1972) sees the villain of the piece as the tourism establishment which has become successful mainly through its "fourpence off" formula (Rivers, 1972: 154). "Tourism can commercialize anything and turn even a personal ceremony into show business" (Rivers, 1972: 162). In Bali, tourism has commercialized religious ceremonies and trance dances (Rivers, 1972: 66), while "in North Africa, the Bedouins have left their flocks to shepherd the tourist, and the ship of the desert has become a pleasure boat" (Rivers, 1972: 164). For the above reasons, Fussell (1979: 34) maintains that the real objects of tourism advertising "are not the Taj Mahal or Mount Etna at sunset, but the Ocean Terminal at Hong Kong", which one waggish travel writer has amusingly christened "The Great Mall of China" (Barry, 1992).

From the manipulations of the tourism industry this line of thinking is quickly extended to the area of human relationships. Fussell (1979: 36), for instance, argues that certain members of the host society, on account of the commoditizing process, soon become transformed into touts, beggars, pimps and whores, while the tourist plays the role of big spender exercising power in deciding what to buy. Papson (1979: 255) goes on to point out that, with ever increasing numbers of visitors, tourists are no longer regarded as recognisable strangers but as unrecognisable foreigners with money. The latter in turn become exploitable commodities and similar transformations occur within the residential community. Crick (1989b) even seems to go so far as to suggest that, with a host dislike of hippies, lumpen Bohemia and Eurobums, who somehow personify conspicuous under-consumption, locals appear to be conspiring with the tourism establishment in extending a greater welcome to the affluent.

Since the first version of Greenwood's (1977) memorable essay on tourism as cultural commoditization, so appropriately entitled *culture by the pound*, his description of the Alarde ceremony of Fuenterrabia, and its subsequent loss of meaning due to commercialisation, both advocates and critics of this working hypothesis have shifted their positions into the larger "authenticity debate" in tourism.

The former point of view, originally articulated by Boorstin (1987), emphasises that tourism stages "pseudo events" for its gullible clients, thereby rendering them "cultural dopes". English (1986: 53-54) seems to identify with this position in his examination of North-South tourism when he states:

> Songs and dances, once restricted to the temple or holy days, are now repeated every day in restaurants and night clubs. Performances are condensed or altered to suit tourist schedules and tastes. Handicraft products are often simplified and mass produced in order to maximise sales to undiscriminating foreigners...The meaning is lost; money is found.

English (1986) additionally believes that, by attaching a price tag to the simplest service, what was once spontaneous and natural has now been replaced by the contrived and commercial. Other supporters of this point of view include such commentators as Britton (1980), Crick (1989a), Erisman (1983), Fussell (1979), Papson (1979), Rivers (1972) and Turner and Ash (1975).

The most notable reaction to the foregoing came from MacCannell (1976), who stressed that some tourists, far from being superficial nitwits, actually seek authentic experiences in other cultures. Their quest, however, is largely thwarted by the tourism industry itself. Cohen (1979), while agreeing with MacCannell that *some* tourists (especially the more alienated experimental, experiental and existential types) look for authenticity in the Other, argues that those in the recreational and diversionary modes pay little attention to such matters. Instead, they may actually go along with staged authenticity and enjoy the post-modern contrived experience (cf.

Moscardo and Pearce, 1986). Consequently, Cohen (1988) maintains that it does not necessarily follow that cultural performances and artefacts lose their meaning simply because they have become objects of commercial exchange. In fact, they can often take on new meanings just as the totally artificial attractions of hyper-reality - such as Disney World - can become part of authentic American (late-modern) culture (Cohen, 1993). Similarly, Selwyn (1992a; 1992b) believes that it is altogether too facile to maintain that there is a one-to-one correspondence between inauthenticity and commoditization.

What then of members of Third World island micro-state societies, such as Barbados? What can be inferred about their host attitudes towards commoditization as a facet of the post-modern condition in the guest? At the political level, a number of ministerial utterances in Barbados would seem to indicate both an awareness of, and desire to, maximise the patterns of consumption associated with international tourism. For instance, the Barbadian Minister of Community Development, in his feature address at the 1992 awards ceremony for the National Independence Festival of Creative Arts, reportedly:

> told the participants that Barbadian culture was to be marketed as a means of earning foreign exchange. The Minister said that it was important to recognise that a country's artistic heritage and its culture in general are actually marketable commodities (*Sunday Advocate*, 1992).

The following year, at the gala opening of the same event, the acting Minister of Tourism, Culture and International Transport:

> reaffirmed government's commitment to the development of the arts and a cultural industry

and went on to say that:

> Barbados has the potential to be marketed as a country in which the arts flourish *(Barbados Advocate*, 1993e).

Then again, the Permanent Secretary in the Ministry of Community Development and Culture, at the opening of the Oistins Fish Festival, and in reference to cultural festivals in general, was heard to say:

> it is known that there is money to be made in this area (Sunday Advocate, 1993a).

One month later, during a seminar "aimed at encouraging young people to develop careers in culture", the Chief Labour Officer told participants that they had "got to treat skills as a business". In an interesting reference to the importance of advertising, the speaker noted:

> a lot of artists knock around with ideas in their heads, but they will not be able to sustain them if they do not grasp marketing (Sunday Advocate, 1993d).

That premier cultural events can be successfully commercialised may be gauged from a report on the sponsorship of the 1993 *Crop Over Festival*. Historically, "Crop Over" derived from the days of plantation slavery where the end of the sugar season was celebrated with a procession of decorated carts bringing in the last load of cane, followed by a party in the mill yard. The accompanying ceremonial burning of an effigy of Mr Harding symbolised both the cruelty of the overseers and the hard times of unemployment lying ahead (Spencer, 1974). Today, Crop Over has become a huge commercial enterprise spanning three weeks, and is targeted at tourists to coincide with a customary off-peak period. Apart from a watered down version of the delivery of the canes and a sponsored "coronation" of the king and queen of the crop, all that remains of the original is a parade of decorated carts, but the latter event is so heavily commercialised with carnivalesque floats bearing the names of leading companies, that few would remember its traditional significance. Sponsored calypso competitions and the "jump up" of *Kadooment* have replaced the earlier and simpler celebrations. According to the report:

> Mobil and Texaco are now in their fourth year of sponsoring the promenade and culture village respectively, while Shell and the Barbados National Bank are in their third year of being associated with the decorated cart parade and folk concert. Solar Dynamics will this year be donating water heaters to the king and queen of the crop.

In commenting on the foregoing, the Chief Cultural Officer of the National Cultural Foundation observed that:

> Corporate Barbados benefits enormously from Crop Over; it is something that is mutually beneficial...We would like to see corporate Barbados advertising, using for example the decorated cart parade as a medium for advertising (*Barbados Advocate*, 1993a).

That money can be made from the Crop Over Festival, not only in Barbados, but also via cable television for post-modern punters in New York, can be gauged from the advertisements placed in overseas newspapers.

A similar commoditization process may be observed in relation to Barbados' leading sport - cricket - which is being increasingly used by the tourism authorities to bring visitors to the island. At one time this cultural activity was almost a religion with the legendary three W's (Weekes, Walcott and Worrell) achieving world status, and the Kensington Oval being regarded as the headquarters of the game in the region (Potter and Dann, 1987: xxi-xxii). Today, with Barbados' fall from the position where it regularly used to contribute over half of the West Indies' eleven, commercial interests have begun to take over. Thus one reads that Trinidadian batting sensation - Brian Lara - (incidently seen in Barabdos as a mere shadow of the Barbadian Sir Garfield 'Gary' Sobers), comes to Barbados of all places to promote Chefette restaurants and Cola Cola. According to the newspaper account:

> Space Aliens, Polar Bears, Digging Dogs, Brian Lara. This is the new campaign for Coca Cola around the world "Always Coca Cola" (*Sunday Advocate*, 1993c).

In the words of a spokesman for the multi-national company:

> This is a good example of utilising Lara at a Caribbean-wide level. His youthfulness is in keeping with the target market of the product ... His performance has made him a household word in cricket and his image is in keeping with the vibrant image of Coca Cola (*Barbados Advocate*, 1993b).

Arguably, one sure sign that culture has become commoditized through tourism is the ever-increasing allusion to tourism as a product. However, the connection between the offering and the post-modern consumer is perhaps best summed up by Executive Secretary of the Caribbean Tourism Organization in his address to the Old Harrisonian Society in Barbados, when he told his audience that:

> Caribbean tourism is a First World product in a Third World region.

Interestingly, he went on to tell his élite audience:

> Old concepts of sovereignty are changing for small and large states. In the global economy, attachment to national symbols and trappings of sovereignty for their own sake, will become irrelevant (*Barbados Advocate*, 1993c).

The Barbados Tourism Authority seems to share Holder's view, though maybe for different reasons. After attempting to evoke national consciousness through its slogan "Tourism is our business. Let's play our part", it exploited the idea of a "shopping paradise" in a big way.

According to the island's Trade Commissioner, by the year 2003 Barbados will have 250 duty-free shops (*Barbados Advocate*, 1993f). However, if this estimate is anywhere near correct, and if locals are unable to make purchases in such emporia through lack of access to foreign exchange, it will effectively mean that tourism will have overtaken the lion's share of the retail sector and rendered members of the host society as strangers in their own land.

At the same time, there appears to be little official enthusiasm for the less formal aspects of commercialised tourism. One thinks, for example, of the beach economy and the various vendors who sell everything from black coral, to cocaine and sex. According to one report, it is now even possible for female tourists to "rent-a-dread" (*Sunday Advocate*, 1993b). Yet, as McFarlane (1989: 2,10) points out after interviewing a local beach boy who boasted "I sell to anyone who wants to buy", the typical local reaction was "he wouldn't recognise an honest dollar if it jumped up and bit him".

So has Barbados actually become a commercial paradise where commoditized cultural items are peddled to alienated citizens of the Metropole? Has it become a centre of pseudo events for First World cultural dopes, or do its contrived attractions offer possibilities for new authentic meanings? This brief overview has served to show that the evidence for, and against, these two positions is quite limited and fragmentary. On the one hand, so far there are no theme parks in Barbados; as yet there are no casinos and other Mafia-style operations to demonstrate that the island has arrived at a state of total commercialization. On the other hand, there are indications that attitudinally and behaviourally the process is well under way. In the latter connection it might be worthwhile to heed the prediction of McElroy and de Albuquerque (1993: 5) made in relation to the wider Caribbean:

> With increasing tourist penetration there is rising affluence, infrastructure and facility growth, external consolidation, increased visitor density and crowding and falling length of stay, asset quality and the gradual substitution of man-made [sic] attractions... for degraded natural amenities.

Hawking Heritage

Apart from a reference to Barbados' Crop Over Festival, which was seen to bear scant relationship to its historical predecessor, and to Greenwood's (1977) treatment of the Alarde, so far examples of cultural commoditiz-ation have concentrated on touristic consumption of the present. However, equally important to this process it the capability of post-modernity to focus on the past, or indeed the future. In fact, the most spectacular shrines of hyper-reality today - the worlds of Walt Disney - have old-time Americana standing side by side with Tomorrow Land and the Epcot Center (Cohen, 1986: 14).

In so far as history is concerned, its political appropriation has come to be known as "heritage", so that "to speak of heritage is to speak of politics" and "the discourse of heritage is an inherently political discourse which by ordering the past orders also the present" (Allcock, 1992: 3). Moreover, "heritage is not just that which has come down to us from the past. It is one version of the past; one which potentially competes with other possible versions, but which has come to be sponsored as appropriate and acceptable" (Allcock, 1992: 2). Thus, what constitutes the past and what is to be excluded is to participate in a "social construction of reality which is contested" (Allcock, 1992: 3).

Traditionally, the definition of past situations has been predicated on national interests. However, with the internationalization of contemporary tourism (Lanfant, 1980), many of these territorial interests have become overridden, and with their demise has come the collapse of boundaries and former bases of power. Indeed, the centre of this power-shift has to be understood in terms of the political economy of the tourism industry. Today it is tourism which "provides institutionalised and prestigious forms through which these ideological processes can be mediated (the presentation of a site; the performance). Tourism also supplies ready made the rhetorics of presentation which conjure these politicised views of the past into being" (Allcock, 1992: 15). Nicholson (1984: 216) captures something of the helplessness of the post-modern tourist in the following dialogue between Guy de Chardonnet and Columbine, the chief characters in *Chase the Moon :*

> Guy: Let me give you a little advice. I don't think you should get involved, there's nothing you can do. It would be better if you forgot everything I've just told you and stayed a simple tourist.
> Columbine: It's too late, I am involved...You can't be a tourist all your life.

Selwyn (1992a: 118) in a less literary fashion, makes a complementary point when he explains that in MacCannell (1976):

> the tourist... goes on holiday in order to recreate, frequently with the help of representations from the imagined pre-modern world, the structures which life in the post-modern world has appeared to demolish (cf. Lowenthal, 1985).

Even Crick's (1989b: 40-41) hippies in Sri Lanka are described by school children in Kandy as "trying to escape the pressures of western industrial society".

Other commentators focus on nostalgia in the tourist. Cohen (1985: 299-300), for example, believes that recreational tourists, in attempting to recapture a transcendental Reality (which according to their own world view does not exist), while mournfully regretting the loss of that Reality, only enjoy its playful re-enactment. For Cohen this is nostalgia in the deepest sense, one which is a contemporary version of Eliade's (1968) "nostalgia for paradise". Elsewhere, he continues the argument:

> The modern world is progressive and calculable but also prosaic and in many respects unattractive. Hence modern man looks back nostalgically to "the beginnings", to a pristine world when nature was "untouched" and life simple or "primitive" (Cohen, 1986: 13).

Thus, for Cohen (1986: 14), whether one is talking about archaeological sites, souvenirs or even entire destinations, there are the recurring themes of "the older the better" and "the new destroys the old". However, recognising that the ancient is not always available to the modern, the tourism industry can feed off the nostalgia in its clients by recreating the past, and it is about this turning of events, attractions and destinations into heritage that authors such as Urry (1990) have so eloquently written.

As far as the wider Caribbean is concerned, McElroy and de Albuquerque (1993: 15) highlight a related strategy undertaken by Bermuda's tourist authorities. In order to emphasise quality over quantity, and to enhance sustainability in the light of declining arrivals, they have turned to "the restoration of historical buildings as tourist facilities and cultural re-enactment and pageants for visitor enjoyment."

In Barbados, where the "Little England" image is still exploited (Wilson, 1993), we have noted elsewhere (Dann and Potter, 1997) that:

> With essentially First World post-modern visitors unable to handle the pace of life and future shock (Toffler, 1970) in their own countries, being provided with selective glimpses of the simple life of yesteryear in Third World destinations like Barbados, may be one of the most lucrative and exploitable trends in contemporary tourism.

On that occasion we gave instances of maximising the nostalgia factor from the Barbados National Trust's "Open House Programme" and "heritage passport" (which offer discounts to various historic sites, including a number of plantation houses, military posts, the museum and synagogue). By way of endorsement, travel writer Rick Sylvain (1993) tells his Canadian audience that:

> heritage passport holders discover a Barbados endowed with history and

texture, tropical delights and architectural riches.

He continues with some examples:

> From restored Gun Hill signal station, dating from 1818, you get the same panoramic view flagmen had reporting the approach of enemy ships. Above the Scottish-like hills of the east coast is Morgan Lewis sugar mill, the only complete windmill left in the Caribbean.

According to English photographer Ken Keene (after his twenty-eighth visit to the island and some 16,000 pictures to his credit), the best way to sell Barbados is to concentrate on the past. As he tells a *Barbados Advocate* (1993g) reporter: "Market its magic and its historical side".

However, in "hawking heritage" it is not simply a question of including historical sites in a diversified tourist product. Rather it is their selection and manipulation for commercial gain. Thus, for instance:

> Sunbury Plantation House is over *300 years old*; a lived in Estate House featuring fine Barbados *mahogany antiques, original old prints* and a unique private collection of *elegant horse-drawn carriages*. Sunbury creates a vivid impression of *life on a sugar plantation in the 18th and 19th centuries* which visitors are very welcome to see (italics added) (Barbados Holiday Map, 1991).

In this piece of promotion the markers of antiquity are quickly signposted in the italicised expressions. Visitors are subsequently "invited" (for a small fee, of course) to take afternoon tea at a 200 year old mahogany table. The logo of this stately home shows an elegantly dressed lady, complete with parasol, descending from a carriage where the horses are being tended by a groom. The reader (potential tourist) is persuaded to sample some of this old-time luxury (reminiscent of Gottlieb's (1982) king/queen for a day). There are no references to the horrors of plantation slavery which have made such opulent living possible.

A similar historical amnesia, which filters out the less savoury incidents of the past by exclusively focusing on the wholesomeness of bygone experiences (a reasonable enough definition of nostalgia), may be found in the tours conducted over the Mount Gay Rum Refinery (Mount Gay, 1992). Here patrons are informed that researchers have established that:

> Mount Gay is the oldest distillery in Barbados and perhaps the world... Born in 1663, Mount Gay rum draws its unique and distinctive character from the island's natural resources and the ingenuity of its pioneering settlers.

The twenty-five minute tour of the stills, cellars and ageing casks, ends with a sampling of the product "while relaxing on the verandah of a traditional Barbadian house that overlooks the Caribbean Sea" (Mount Gay, 1992). Apart from all the historical scholarship (as a guarantee of authenticity), and the company's seeming endeavour to demonstrate its potential candidacy for the *Guiness Book of Records*, there is no mention of the appalling conditions under which rum was originally produced, no allusion to the deleterious effects on the population of this "kill devil", and of course no reference to the contemporary reality that Mount Gay is now owned and managed by the French multinational Remy Martin.

From stately homes and tours it is but one step to the staging of history through dramatic productions. One of these shows is called *"1627 and all that"* (a clear borrowing from "1066 and all that"), though referring to Barbados' initial colonisation by the English. Nevertheless, this historic landmark is spoken of as "100 per cent Bajan and 100 per cent entertaining...a colourful cultural extravaganza of Bajan folklore and music told in music and dance" (*Visitor*, 1993a: 3). Another show - *"The Plantation Tropical Spectacular"* - is described in its promotion as "a potent blend of flamboyant folklore and fantasy. From the early days of pirates to the carnival of the underwater world" (*Visitor*, 1993b: 5).

Yet another show, called *"Barbados by Night"* and featured in the same advertisement, invites patrons to "see the cultures of the Caribbean as influenced by the Spanish, French and African *Settlers*. Plus fire-eating, flaming limbo and steelband (italics added)". Apart from the historically inaccurate bestowal of *settler* status on Africans, and their unique cultural contribution being spoken of solely in showbiz terms, it is also clear from these examples that flamboyance and fantasy have removed an authentic portrayal of life on a plantation and replaced it with pirates and carnival. It is also unlikely in the extreme, of course, that female slaves would have appeared in two-piece sequinned costumes with exotic feather fashioned head-dresses, or that their male counterparts would have spent their days solely engaged in fire eating, limbo dancing and the production of steel band music (which incidentally originated in Trinidad and at a much later period). Equally improbable is the realisation that authentic portrayals would be so attractive to tourists or make so much money as the hawking of heritage via the selective staging of the past.

Even more recently, however, there is some evidence to show that, not only is nostalgia being evoked in the guest, but also in the host. In a double-feature advertisement (*Sunday Sun*, 1993) entitled "1993 Heritage Festival", Barbadians were invited to participate in a weekend of "old-time fun". On the Saturday they could go to the Museum and witness a fine craft exposition, while on the Sunday they could patronise a "heritage fair"

- "head out and relive an old time Bajan exhibition" at Hopefield Plantation. When the latter was promoted in the *Barbados Advocate* (1993g), the message was as follows:-

> Recapture the atmosphere of Barbados before the 60's with stiltmen, tuk band, Barbados Landship, steel band music, warri competition, donkey cart rides, old time games and storytelling. Enjoy traditional Bajan fare: puddin 'n' souse, fishcakes and bakes, old-fashioned sweeties, snowballs, conkies, toffee apples, sweetbread, coconut bread, cane juice, mauby, and more...

A subsequent newspaper article on the Barbados National Trust event (Best, 1993) described it as "truly nostalgic":

> The fair was a true reminder of those bygone days when such events were held, where families and courting couples came out to enjoy whatever activities were in store.

And the reason for its success?

> Today, it is not often that one can find a pleasant yet different way to spend the afternoon outdoors.

While the author did not elaborate on why exactly today it is difficult for residents of Barbados to enjoy themselves in the same way as yesteryear (on account perhaps of recession, unemployment, crime, or whatever), it is clear that a dip into the pleasant memories of the past can serve to obliterate the grim realities of the present. In so doing, it may be noted, Barbadians are not unlike the visitors they seek to attract, and for precisely the same motive. By entering a bygone era themselves, they are therefore attitudinally and behaviourally opening up to the sort of tourism on which post-modernism thrives.

Playful Placenessness

In a post-modern world of pastiche, populism and kitsch, where there is a bombardment of media generated images and where the consumption of signs has greater value than their use (Lash, 1991: 36-44), clearly tourism, as emblematic of this ethos, stands to gain through its allied predilection for theme parks and the museumization of culture (MacCannell, 1976; Urry, 1990). In the doing, tourism feeds off the playfulness of its devotees and the placelessness (Relph, 1983) of its staged events.

As far as the ludic and liminoid tendencies of recreational tourists are concerned (Lett, 1983; Cohen, 1985; Graburn, 1992), an "as if" attitude is adopted towards the authenticity of their experiences. Just as the contemporary theologian can playfully treat God "as if" he were dead, so too can the post-modern tourist playfully engage in reconstructed reality "as if" it were genuine, without actually being taken in by its orchestrated performance (Cohen, 1982). Thus, the secular pilgrims to Walt Disney World demonstrate their inherent playfulness by not only knowing the narrative, but also in recognising that attractions are illusions based on trick effects (Moore, 1980). According to Cohen (1993: 13):

> attractions are also a "show"; they therefore bespeak the tourist's predisposition for playfulness, his [sic] readiness ludically to accept "contrived" attractions as if they were real. The predisposition becomes culturally sanctioned by the post-modern ethos.

Later he argues:

> this concern with the enjoyability of the surface appearance of the attraction, rather than its "reality" or "authenticity", makes it possible for the variety of "contrived" attractions to flourish (Cohen, 1993: 15).

It is for this reason that Moore (1980: 207) maintains that play has today supplanted religious ritual, and that Crick (n.d: 2) believes:

> in our post-structuralist phase, with all its ludic overtones, tourism, with its inevitable playful regressive "as if" component, surely has an honoured place.

Interestingly, also, Crick (1985) argues that anthropologists who study tourists are so playful themselves that they are virtually indistinguishable from the subjects of their research. While not agreeing with Crick, Errington and Gewertz (1989: 46) nevertheless maintain that:

> tourists are essentially unilinear evolutionists who find the world filled with chiefs and witch doctors, and their self-referential tales are based on - indeed require - partial, simplified and other completely erroneous information.

Just as playfulness in the tourist leads to the manufacture of contrived experiences, so correspondingly does the importance of place diminish. After all, if the hyper-real becomes better than the real through sophisticated technology, it hardly matters any more where the staged production is located, or (apart from climatic considerations) whether Walt Disney's kingdoms are to be found in Florida, California, Paris or Tokyo. The fact that the "Many Worlds" theme park of Sentoza is situated in Singapore, along with its rain forest, ruined city, lost civilization and dragon walk, is somehow less relevant than the realisation that it has surrendered geographical position to fiction parading as truth (Cohen, 1993: 7-8). Maybe this is also why Club Med can establish a tropical paradise right in the heart of Vienna (Cazes, 1987: 13). Similarly:

> today's quintessential tourist state is Florida where the natural wonders of the Everglades and the Keys are totally swamped by the man-made attractions of Disney World, sea aquariums, parrot and orchid jungles, and the circus showcase of Ringley Bros and the Barnum and Bailey (Turner, 1976: 20).

The object of the tourist gaze (Urry, 1990) has become a spectacle, a theatre, a "spielhaus" (Buck, 1978: 225):

> The places of the glossy brochure of the travel industry do not exist, the destinations are not real places, and the people pictured are false (Crick, 1989a: 225).

Although no areas in Barbados are totally given over to thematization, nevertheless one can find ludic centres which may contribute to a sense of placelessness. As we have noted already, there are shows which make a playful spectacle of Caribbean history, along with the staging of events such as Crop Over which bear only a remote resemblance to their origins.

But it can equally be argued that such flexible spacelessness is further bolstering the advantages and pulls of the existing urban nexus in a small island developing state such as Barbados. Then too, there are local cruises to be had aboard the *Jolly Roger* or *Bajan Queen* crafts. While the former plays upon the days of buccanneers, patch-eyed pirates, parrots, skulls and crossbones, and even stages walk the plank mock weddings for inebriated tourists, the latter cruise can take its equally drunken passengers back to the era of the Mississippi tramp steamer. Such pleasure trips are not unique to Barbados. Indeed, they can be found in other parts of the Caribbean and elsewhere. Even Malta contributes to playful placelessness with its own version of Captain Morgan Cruises (1993), and, just to top it off, during the English language commentary on the Malta Experience, visitors are regaled with West Indian calypso.

Just as hosts can be acculturated to the out-of-time nostalgia of their guests, so too can they be prepared for parallel out-of-place experiences. The latest of these fantasies to be enacted in Barbados was "Christmas Wonderland" in which a leading car retailing company had transformed its showrooms into "an unforgettable experience of sight and sound in the true Christmas spirit". Boasting 50,000 lights, 65 Christmas trees, 25 animated characters and a life-size creche, locals could "enjoy polaroid photo sessions with a live Santa Claus, stroll through Santa's workshop and living room, (and on through)...a clock shop and miniature village". Additionally, patrons were "entertained with a live choir every night, while taking...sleigh rides and sampling a variety of Bajan goodies and drinks" (*Barbados Advocate*, 1993h). Arguably, it would only have been the last ingredients which would have reminded customers where they really were. Otherwise they would have been completely immersed in a world of Santa and Rudolph, where even the church was that of Bethlehem (*Sunday Advocate*, 1993f). Yet for many Bajans, whose only experience of snow or cold had been obtained vicariously via television or the cinema, Christmas Wonderland would have opened the doors to "as if" simulations. By submerging themselves in make believe through play, they were somehow both anticipating and entering the ludic world of the recreational tourist.

Blurring Boundaries

Clearly, a sense of placelessness is closely associated with a blurring of boundaries, a condition which in turn can be identified with the processes of post-modernization and globalization. In this connection, Lash (1991: 11-12, 26) speaks of a loss of autonomy of cultural spheres, a de-differentiation of high and low culture, a blurring of the distinction between the cultural and the social, reality and representation, representation and advertisement. In architecture, too, modern rational forms have given way to post-modern ornamental play and pastiche more reminiscent of the medieval labyrinth and a return to the vernacular (Lash, 1991: 33).

It was MacCannell (1976) who first introduced the idea of the "tourist space" in order to distinguish it from the ordinary lives of destination peoples. With this concept came the whole authenticity debate and such issues as to what constituted a genuine experience and whether or not the tourism establishment were interested in providing it. Subsequently, Buck (1978) applied the notion of "boundary maintenance" to tourism. At that time he reckoned that it was useful for the industry, and patently, in so far as visited peoples were concerned, to maintain boundaries. And so it was that the private lives of the Old Order Amish were preserved from prying tourists by a system of replicas and staged attractions. After a while it would seem as though clients of the Dutch Wonderland, Wax Museum, Amish Homestead, Kitchen Kettle, Dutchland Tour and Farmer's Market, were hardly able to distinguish their experiences from the real thing. Arguing from a geographical perspective, and in relation to Caribbean destinations, Husbands (1986) subsequently examined the allied question of competition for space between tourists and residents. There he maintained that mature destinations like Barbados suffered less from this problem than islands which were unaccustomed to dealing with outsiders.

Certainly in Barbados there are well-established barriers of social class, area of residence, party political affiliation, race and gender (Potter and Dann, 1987). It is also true to say that the economic dividing lines are well drawn between the rich and poor. Yet increasingly, there are several indications that at the cultural level, less fragmentation is in evidence. Whereas formerly "the arts" constituted the prerogative of colonials and local élites, now there is a far greater spirit of democratisation in creative writing, painting, theatre, song, dance and music. The distinction between high and low culture has all but evaporated.

It is this cultural levelling among residents of Barbados which has its touristic analogues and which somehow provides coping mechanisms of supply for post-modern demand. Dining out, for instance, was at one time considered as the high point of a bank holiday excursion, an *al fresco* affair synonymous with consuming vast amounts of local food and rum. Today there are restaurants specialising in Chinese, Italian, French, Mexican, American and Indian fare, and gala nights sponsored by the Barbados Hotel Association somehow manage to combine a selection of all these different national dishes into one enormous buffet. Additionally, several wine and champagne bars have sprung up as alternatives to the less sophisticated institution of the local rum shop. Then, too, names which Barbadians give their children are also indicative of an emerging cultural pluralism. Once these were restricted to those of saints and biblical characters. Now they range over entire continents and feature the appellations of Indian princes, African folk heroes, American pop stars and European entertainers.

Such diversity is also reflected in clothing where a number of fashion experts have emerged to cater to, as well as to create, new and exciting designs. While many of their products resemble works of art which playfully combine motifs from around the world, several youngsters prefer to don apparel from the United States with their designer jeans and LA gear. Indeed, the proliferation of tee-shirts, and an equally rich array of extra-regional messages would provide sufficient material for a study on its own. Hairstyles, too, demonstrate today a far greater diversity than the earlier predominant reliance on the "afro". There is also a parallel interest in body care, as a growing number of cosmetologists, skin products and beauty treatments make their inroads on to the local scene.

The sources of inspiration for such an admixture of personal tastes and lifestyles are many. In particular, they stem from the electronic media, videos, magazines and advertising, where on a daily basis Barbadians are bombarded with images from around the globe. Some also originate from individual travels undertaken by Bajans to the urban shopping meccas of Miami, New York, Toronto and London, as well as from the so-called "demonstration effect" of having many tourists in their midst from these metropolitan centres of consumption (see Potter, 1993c; also chapter 3). The influence of families and friends who have emigrated northwards, who visit and are visited, and who keep in touch with those back home, also needs to be added to the equation of post-modern "to-ing" and "fro-ing".

For these reasons it is often difficult to discern boundaries between host and guest or to determine whether culture trickles down or up in a situation where the distinctions between the centre and the periphery have become increasingly blurred. One example of such fuzziness may be gleaned from an account by "Cynthia" (1993) concerning the Oistins Fish Festival:

> The street party reminded me of Camden High Street in London on a Saturday morning, only much better. In Camden you'll find people selling everything from pirated records to Chinese jewellery, while at Oistins I saw people selling everything from goofy looking inflatable animals to chances to win groceries.

Cynthia goes on to describe a "dub stall, for want of a better name" where people were "bogling, dipping, rocking and butterflying". She also refers to the Monday crowd as a "fashion conscious and snazzily dressed posse". While some were attired in Easter Sunday best, others were casual "à la day in the garden". What took her attention, however, were those "dressed to kill in these ripped up and patched suits", referred to as "absolutely brilliant works of art", but which she learned were called "Major Damage".

On the other hand, a blurring of touristic boundaries may constitute an unsettling experience, particularly for the older generation. Thus, for instance, former Minister of Tourism, Peter Morgan (1992), describes the touristic transformation process as:

> (a)...proliferation of signs and banners which have turned St Lawrence Gap, Rockley and Holetown into cheap replicas of Miami Beach, Atlantic City or Blackpool.

Interestingly, in making his derogatory comparisons, Morgan employs a post-modern emphasis on "signs", "banners" and "replicas", as well as stressing the urban base of these post-modern touristic developments. By referring to St Lawrence Gap, he also, wittingly or otherwise, highlights one coastal area of Barbados which perhaps more than any other epitomises the post-modern urban ethos and the blurring of boundaries between the visitor and the visited. Not only is the St Lawrence area given over to a wide variety of hotels, ethnic restaurants and entertainment centres, but also it is home to a "chattel village". Here the traditional vernacular wooden house has been uprooted from its natural habitat and used in a series of retail outlets under such names as "Fat Willy's", "Lazy Days in the Gap", "Bits 'n Bobs", and even "Beach Bum Exciting Tropical Beach Wear" (Miller and Miller, 1991: 55). In the "chattel village", the back-stage region of authenticity (cf. MacCannell, 1976) has been moved front-stage for commercial purposes. Whether or not the recreational tourists are aware of the deception, or whether they enjoy the contrived "as if" experience, is perhaps less important than the realisation that this is yet another example of "urban-based placelessness" and a further obliteration of the lines of demarcation between tourists and residents. At the same time, the creation of an artificial village in an urban context for post-modern consumers, does at least have the advantage of helping to protect the real rural and semi-rural villages of the inhabitants from intruders.

A similar strategy has been employed by a number of hotels. At Marriott's Sam Lord's Castle, for instance, guests are served the elements of a buffet dinner from a "rum shop" and some "chattel houses" which have been specifically constructed on the hotel premises. For this reason, tourists patronising this establishment may no longer find it necessary to seek out the genuine article, particularly when the distinction between the real and the inauthentic has become temporarily suspended and where the vernacular has been appropriated. In the Hilton Hotel (incidentally a majority-owned government property), a playful collage of local architectural forms has been constructed within the building to yield a verandah restaurant. While patrons are unlikely to confuse this pastiche with the reality outside the confines of the enclave resort, it is important to recognise that in many ways the hyper-real bricolage of replicas is not only better assembled, but appears to be in superior condition to most of the local timber housing stock. If guests can therefore take in some local culture amid the cocoon-like safety of surrounding "America", without having to experience the dangers associated with strangerhood, a little blurring of boundaries actually redounds to their benefit.

In the other accommodation establishment where the Barbados Government held a financial interest - Heywoods - management had even found a way of bringing what has become a distinct cultural institution into its own dining facilities (Nation, 1993). Baxter's Road in Bridgetown ("the street that never sleeps"), especially during Crop Over, is the scene of all night partying, rum shops, fish fries, dancing, and the general mingling of tourists and locals. Now, however, it is quite possible to enjoy a simulated experience without the hassle of stepping out of the hotel. The worlds of the indigenous and the touristic have blended into one.

Conclusion

In identifying and discussing five themes from the literature, "peddling paradise", "commoditizing culture", "hawking heritage", "playful placelessness" and "blurring boundaries", we have attempted to show by means of a number of examples that there are several local responses to the presumed condition of post-modernity which characterises many tourists. We are certainly not implying that such an ethos is to be found in all tourists. Rather we agree with Cohen (1979, 1993) that it is more evident in the recreational and diversionary varieties attracted by "sun, sea and sex" than in more serious types of travellers, wanderers, drifters, explorers, adventurers, nomads and pilgrims.

Barbados, as we have seen, is a "mature" destination which is primarily geared to the accommodation of mass tourists in luxury hotels and resorts, many of which are owned and operated by multi-national companies intimately tied into the ideologies and best practices of the so-called "new world order". The majority of hotels are located on the coastal urban corridor. Most of Barbados' clients are drawn from First World, Third Wave, societies - the United States, Canada and Europe – urban/metropolitan societies already well aware and daily affected by a post-modern ethic of consumption. By thematising this cultural condition, and by examining local strategies for catering to such demand, we are additionally suggesting that the effectiveness of indigenous response is predicated on transformations which have already occurred or which are currently taking place in the host society. Some of these changes are due to tourism itself (as both a recipient and participant activity). Others are linked to the forces of post-modern acculturation (the mass media, advertising, public opinion), which in many senses have come to replace the more traditional socialising agents of family, church and school.

The themes under which we have treated these issues are thus just as applicable to the host as to the guest. They are not intended to be exhaustive of the post-modern condition. We could, for instance, have devoted space to an examination of images, signs, critiques of scientific realism and metanarratives, the crises of legitimation and rationalism, even the radical conceptualisation of language (cf. Denzin, 1986). Rather, the themes have been selected as being the most appropriate to contemporary tourism, to the extent that they have already been highlighted in the writings of others.

Nor, for that matter, are these themes to be considered as mutually exclusive. Indeed, in many cases it has been indicated that they lead into, and feed off, each other by both corollary and association. Paradise becomes contrived by being locked into consumerism. The pristine, the natural and the ancient become artificially assembled, preserved or created, or else turned into commoditized products for the nostalgia industry. The playful "as if" experiences of the tourist seeking paradise, heritage and authenticity, are matched by ludic centres and staged events. The realisation that the latter could be anywhere is conducive to an examination of placelessness, and placelessness in turn is characterised by a blurring of boundaries. As a matter of fact, it is this very inter-linkage which lends itself to the pastiche and de-differentiation so emblematic of post-modernity. Thus, while treating the above themes as separate for analytical purposes, we would be the first to acknowledge that they are very much an amalgam of forms which constitute the combined characteristics of contemporary life.

Finally, it should be made clear that nowhere are we suggesting that loss of meaning it to be automatically attributed to the processes outlined above. Although there is certainly some meaninglessness associated with checking out of Sainsbury's and into Barbados with the mere flash of a switch card, and all within the same High Street (Selwyn, 1992b: 18), we prefer to emphasise that it is rather a change of meaning which is taking place among visitors, the visited and even those who comment on such matters (Crick, 1985). It is our hope that, within the limits of this presentation, our line of reasoning can act as a catalyst and be extended and continued, so that we will be in a better position to evaluate Crick's (1989a: 333) observation that:

> We now live in a post-Marxian world of the political economy of the sign; the emphasis has shifted away from production itself to image, advertising and consumption. We are now interested in what Baudrillard has termed "the mirror of production", and tourism, being so much a matter of leisure, consumption and image, is an essentially (post) modern activity.

Whilst much of this image creation, advertising and consumption is relatively flexible in the post-modern world, in the Caribbean it is increasingly coming to be associated with the linear coastal tourist zones - the very zones which were originally etched out by mercantilism and early capitalism. Thus, it is a short step to argue that the condition and realities of post-modernity in the Caribbean are intimately associated with an extending urban-quaternary economic zone, via its direct association with tourism and related urban-based tertiary/quaternary activities.

5 Globalisation, Environment and Urban Development in the Caribbean

Introduction

As elsewhere, there is a growing public and official awareness of the pressing nature of environmental and ecological issues in the Caribbean region. Such concern is witnessed in a number of different ways, not least in the attendance of Caribbean representatives at the Earth Summit or United Nations Conference on Environment and Development, which was held in Brazil in June 1992, and the associated emergence of Agenda 21. Many also participated in the deliberations of May 1994, when Barbados acted as host for the United Nations conference on sustainable development strategies in small island states (SIDs). This chapter widens out the account presented in chapter 4, stressing the interactions existing between global trends and the development-environment interface. In this sense the focus is generic rather than being exclusively urban-based. However, issues relating to the role of urban areas and urbanisation in perceptions of land, the effects of global climatic change on urban conditions are specifically highlighted in this chapter.

In popular terms, environmental awareness was shown by the Peoples' Parks which sprang-up in many areas of Barbados at the end of the 1980s, testifying to spontaneous efforts on the part of the public to beautify the environment. It is also demonstrated with the appearance of various environmental associations, "green teams" and the various conservation strategies undertaken by the Barbados National Trust and other NGOs. The longstanding efforts of West Indians with respect to the upkeep of their popular dwellings (Potter, 1992c, 1995), and in relation to family land (Barrow, 1992), also attest to this recognition of the salience of the environment and the need to promote environmental sustainability.

Caribbean Views on Environment: An Initial Empirical Perspective

David Lowenthal, who at that time was working for the American Geographical Society, published a paper under the title 'Caribbean Views of Caribbean Land' in the *Canadian Geographer* in 1961. In it he presented the argument that in the Caribbean, "the association ... between man (sic) and land" is "entirely commercial", and that, "In the Caribbean, and notably its dependent areas, land is seldom mentioned (save by tourists) except as a commodity" (Lowenthal, 1961: 1). Lowenthal went on to observe that the "View that land is valueless save as a commodity persists in many aspects of West Indian life today" (Lowenthal, 1961: 3). Lowenthal argued that such attitudes can only in part be accounted for by the small size and dense population of Caribbean territories, along with the over use of land which is so commonly encountered in the region. Thus, Lowenthal argued that in other areas of the globe, where similar physical geographical realities prevail, for example, the Pacific, the aesthetic, religious, commercial and national significance of land assumes far greater importance.

The reasons given by Lowenthal for what he regarded as the disinclination to adopt non-economic and non-commercial perspectives on land in the Caribbean, basically involve the influence of slavery and colonialism. Thus, his analysis initially stresses that the Caribbean is a place and not a people, due largely to the heterogeneity of the population, which it was argued gives rise to a lack of attachment between Caribbean lands and their inhabitants. Secondly, the fact that land in the region is all too frequently associated in the minds of many with colonialism, is seen as causal. Thus, absentee proprietors and the repatriation of profits have led to many regarding land principally as a machine for making money. Lowenthal, therefore, argued that there is a general lack of attachment to country in the Caribbean, citing perhaps unfairly, the ramshackle, makeshift appearance of many West Indian houses as evidence supporting this. Although the veracity of this particular example must be questioned, Lowenthal more fittingly presented the conclusion that "The productive West Indian soil, as well as the soil itself, are rejected as relics of slavery and shame, unfit for consumption by free men" (Lowenthal, 1961: 4).

The present author undertook a number of cognitive social surveys in order to investigate, in a preliminary manner, the nature of Caribbean views on environment and development (Potter, 1992c). The approach initially involved a repertory grid survey based on the eleven administrative parishes making up Barbados. The names of the administrative parishes were presented to the ten respondents in randomly-selected *triads*, that is, sets of three, in order to attempt to identify the dimensions of appraisal that were being employed in evaluating and comparing areas or environments. The great virtue of the method is that the dimensions of appraisal, or *constructs* are elicited from the respondents, not imposed or pre-specified by the researcher.

The principal salience of the work was that it exemplified, most cogently, the importance which was attached to economic and socio-demographic dimensions of environmental appraisal amongst the respondents, over geographical and infrastructural ones. The respondents produced between four and fourteen constructs, with an average elicitation of 7.1 per individual. Semantically, there appeared to be 28 separate construct-contrast pairs, of which 13 were used by more than one of the respondents (Table 5.1). Only five of the constructs applied to more than half of those interviewed. Turning to the nature of the constructs elicited, they clearly fell into four distinct categories, referring to economic, socio-demographic, physical-environmental and infrastructural conditions. In the final column of Table 5.1, each of the 28 constructs has been designated as falling into the remit of one of these four broad types. The leading six construct-contrast pairs all relate to either economic or socio-demographic circumstances and conditions. It is notable that of all the environmentally-oriented constructs cited, none are appraisive in the sense of dealing with emotional responses to the environment. Rather, where environmentally-oriented constructs do appear, they refer to factual circumstances (hilly-flat, inland-coastal, high rainfall-drier, remote-not remote). Similarly, there are no construct-contrast pairs which reflect cultural-historical categorisations *per se*. A numerical summary of the predominance of economic and socio-demographic construct-types is provided by Table 5.2.

But there is a further fundamental aspect of individual's cognitions of the environment, and this can only be examined by looking at the ways in which the different constructs that make-up the grids pertaining to individuals relate closely to one another. This factorial structuring of repertory grids can be examined by some form of multivariate analysis. In the present case, this was carried out by employing a relatively simple form of non-parametric analysis (see Potter and Coshall, 1984). The results of factor analysing the repertory grids pertaining to the ten respondents are provided in Table 5.3. The respondents are referred to as Subjects A to J. The number of factors elicited varied from two to seven. The first point is that almost all of the factors are coherent, and can easily be interpreted. In virtually every case, the factor accounting for the most variance is a summary measure of relative economic development. Taken together, the development-urbanisation dimension accounts for the first factor in respect of Subjects A, B, C, D and E, or half those interviewed. In the other cases, dimensions reflecting tourism, expensive areas, inland and remote zones

Table 5.1   The construct-contrast pairs elicited, ranked by frequency of citation: perceptions of environment in Barbados

Number	Construct-Contrast	Frequency	Type
1	Agricultural areas – not agriculture	8	Ec
2	Tourist area – non-tourist area	8	Ec
3	Urban area – rural area	7	S-D
4	Densely populated – low population density	7	S-D
5	Developed – less developed	5	Ec
6	High-income housing – low-income housing	3	S-D
7	Hilly area – flat	3	Evt
8	Inland – coastal	3	Evt
9	More shopping facilities – less shopping facilities	3	I
10	Promoting industry – not industrial	3	Ec
11	High rainfall area – drier area	2	Evt
12	Remote areas – not remote	2	Evt
13	Less business/commerce – more business/commerce	2	Ec
14	Clay soils – limestone area	1	Evt
15	People migrating in – more stable	1	S-D
16	Transport problems - accessible	1	I
17	Suburban – not suburban	1	S-D
18	More schools – less schools	1	I
19	Many cars – few cars	1	S-D
20	East coast – other areas	1	Evt
21	Much natural vegetation – less vegetation	1	Evt
22	Becoming more developed – not developing	1	Ec
23	Open to the elements – not open to the elements	1	Evt
24	Historically interesting – not interesting	1	S-D
25	More attractive area – less attractive	1	Evt
26	Socially cut-off – not cut-off	1	S-D
27	Fishing area – not a fishing area	1	Ec
28	Many good roads – few good roads	1	I

Note: Ec – Economic construct type
S-D – Socio-demographic construct type
Evt – Environmental construct type
I – Infrastructural construct type

Table 5.2   Relative importance of the four construct types identified

Construct Type	Frequency of citation	Percentage of citations
Economic (Ec)	28	39.43
Socio-demographic (S-D)	22	30.99
Environmental (Evt)	15	21.13
Infrastructural (I)	6	8.45
Total	71	100.00

are revealed. Notably, in the case of subjects B and C, two components seem to emerge, which serve to stress the relative economic standing of areas. The fact that environmental perceptions focus on economic standing is clearly attested again in this further analysis.

It is equally notable that where physical-environmental construct-contrast pairs form part of the analysis, they tend to float on their own, appearing as single-construct factors. Examples are the single construct factors 'coastal areas' (Subjects A and J), 'flat land' (Subject B), 'less mountainous' (Subject E), 'clay soils' (Subject H), and 'hilly' (Subject I). A further point that emerges is the clear schism between agriculture and tourism which exists in the minds of the respondents. This is brought out in the first or leading factors pertaining to Subjects B, D, F, H and J. In the case of Subject B, for example, the first factor brings together the constructs 'less tourism' (+11) and 'more developed agriculture' (+9). In the case of Subject F, the first factor is made-up in its entirety of the separation between 'tourist areas' (+11) and 'less agriculture' (-11).

It may be argued that in the present-day context, it has been colonial administrations and the plantocracy, together with post-independence governments and local elites that have been the prime agents in promoting the view that land is largely to be considered as an economic entity. Regionally, it is not hard to find strong affinity with locality and great pride in local ways of doing things among members of the general populace. This is true, for example, in respect of the local, traditional house type found in the region, which is still much valued by members of lower income groups within society. It is frequently members of the elite who see the wooden chattel house and the informal sector in general as things of the past, to be eradicated in the quest for so called 'modernity'. This positive view of the local environment amongst individual members of the

Table 5.3   Results of factor analysing the Barbadian repertory grids

Subject A

Factor 1: 76.36 per cent

High population density – small population	+11	
Rural community spirit – *urban way of life*	–11	} urban areas
Facilities available – far away/few attractions	+10	
Less attractive – *more attractive*	–10	

Factor 2:

| Coastal – inland | – | coastal areas |

Subject B

Factor 1: 38.31 per cent

Becoming developed – already developed	+11	
Less tourism – more tourism	+11	
Dense population – *low density*	–10	} less developed areas
East coast – west coast	+9	
Built-up – not built-up	–9	
More developed agriculture – less developed agriculture	+9	

Factor 2: 17.53 per cent

Less developed – more developed	–9	
More subject to elements – less subject to elements	–9	} more developed areas
More interesting historically – less interesting historically	–9	

Factor 3: 12.99 per cent

| *Less business transactions* – more business transactions | +11 | } remote areas (?) |
| *More rainfall* – less rainfall | +9 | |

Factor 4: 7.14 per cent

| Hilly areas – *flat land* | –11 | flat land |

Factor 5: 7.14 per cent

| Inland area – *coastal area* | –11 | coastal area |

Table 5.3 continued

Factor 6:

More natural vegetation – less vegetation	-	more natural vegetation

Subject C

Factor 1: 29.09 per cent

| *Agricultural areas* – less agricultural areas | +8 } | |
| Less developed area – *more developed area* | -8 } | ? |

Factor 2: 20.00 per cent

| *More industrial land use* – less industrial land use | +11 | more industrial land use |

Factor 3: 18.18 per cent

| *More urbanised* – less urbanised | +10 | more urbanised |

Factor 4:

| Less commercial activities – more residential | - | more residential |

Subject D

Factor 1: 65.91 per cent

Tourist areas – not tourist	+11 }	
More agricultural – less agricultural	-9 }	tourist/developed areas
Developed – less developed	+9 }	

Factor 2:

| Non-industrial – industrial | - | non-industrial |

Subject E

Factor 1: 66.23 per cent

Better shopping – poorer shopping	+11 }	
Urban/built-up – few facilities	+10 }	
Dense population – low density population	+10 }	urban/developed areas
More agricultural – *less open pasture*	-10 }	
Less developed – *more developed*	-10 }	

Table 5.3 continued

Factor 2: 10.39 per cent

Less mountainous – more mountainous	+8	less mountainous

Factor 3:

Low-income housing – middle-income housing	-	low-income housing areas

Subject F

Factor 1: 33.33 per cent

Tourist areas – non-tourist areas	+11	} tourist areas
More agricultural – *less agricultural*	-11	

Factor 2: 31.82 per cent

More dense population – less dense population	+11	} densely populated/accessible areas
Remote areas – not remote	-10	

Factor 3: 12.12 per cent

Many good roads – poor roads	+8	many good roads

Factor 4:

Fishing areas – no fishing areas	-	fishing areas

Subject G

Factor 1: 34.09 per cent

More expensive housing - less expensive housing	+10	
Exclusively agricultural areas – *not exclusively agricultural*	-10	} expensive areas with communities
"Cut-off" socially – *not "cut-off" socially*	-10	

Factor 2: 30.68 per cent

Getting more involved in tourism – established tourism	+9	
More urban – rural	+9	} tourist-urban areas
Established tourist areas – not tourist areas	+9	

Factor 3: 11.36 per cent

People migrate in to build homes – stable areas	+10	people migrate in to build homes

Table 5.3 continued

Factor 4:

More densely populated – low population density	-	more densely populated

Subject H

Factor 1: 74.55 per cent

Low population density – high population density	+11	
More agricultural – less agricultural	+10	agricultural-rural areas
Rural areas – commuter zone/non-rural	+10	
Economy tourist oriented – *economy not tourist oriented*	–10	

Factor 2:

Clay soils – limestone areas	-	clay soils

Subject I

Factor 1: 45.45

High rainfall – less rainfall	+11	inland areas
Lower population density – high population density	+9	

Factor 2: 20.45 per cent

Hilly/higher land – rolling land	+9	hilly/higher land

Factor 3:

Developed – less developed	-	developed areas

Subject J

Factor 1: 29.37 per cent

Remote/country – not remote	+11	
Less shopping facilities – more/better shopping	+11	remote areas/poor facilities
Transport difficult – accessible	+10	
More schools – *less schools*	–10	

Factor 2: 20.98 per cent

More agricultural – less agricultural	+10	
High status housing areas – *low status housing area*	–10	agricultural areas
Tourist areas *less of a tourist area*	–10	

Table 5.3 continued

Factor 3: 13.99 per cent		
Suburban – *rural/urban*	-10	
Many cars – *few cars*	-10	} ?
Factor 4: 6.99 per cent		
Rural – *urban*	-10	urban
Factor 5: 6.99 per cent		
Promoting tourism – *not promoting tourism*	-10	not promoting tourism
Factor 6: 6.29 per cent		
No beaches/inland – *coastal/beaches*	-9	coastal/beaches
Factor 7:		
Promoting industry – not promoting industry	-	promoting industry

community is witnessed in the importance that has always been attached to family lands, a point noted by Lowenthal.

But on the other hand, Lowenthal wrote of the "Readiness of individuals and governments to sell and lease lands to foreign companies, ... (in) the easy faith that the highest income per acre is the greatest national good". The accusing finger surely has to be pointed at the lure of the fast buck, so that principles of economic expediency have continued to rule. Hence, the uncritical adoption of the development strategies of More Developed Countries in the region, embracing modernisation and trickle-down effects, whilst little attention has been devoted to agriculture and to the provision of basic needs (Downes, 1980). At the same time, tourism has come to be conflated with the urban, and contrasted with the rural and agricultural.

In the Caribbean, land has always been a vehicle for short-term profit, not a resource to be husbanded for future generations (Richardson, 1992). Further, economic benefits from Caribbean environments have primarily accrued to outsiders. This is perhaps nowhere better attested than in the realm of tourism. For example, as noted in chapter 3, it is estimated that 70 per cent of all tourist expenditure in the Caribbean is repatriated, primarily as a result of the foreign ownership of hotels and the importation of foodstuffs. In Barbados, 75 per cent of Grade I hotels are foreign owned (Potter, 1983). In several instances, whole islands have been sold to private concerns. Thus, Colin Tennant, now Lord Glenconner, purchased the island of Mustique for £45,000 in 1957 (Miller, 1992). In other words, while there may be a rising tide of environmental consciousness in the Caribbean, this is unlikely to be transformed into increasingly sustainable forms of development in nations where land has so often in the past been sold to the highest bidder, and where states are desperate to 'modernise'. And this is in a world which is increasingly referred to as being influenced by 'post-modern' trends, as elaborated in the previous chapter. This makes for some early conclusions that are more generally applicable than to the Caribbean alone. First, sound development involves maintaining a careful balance between indigenous peoples and their environments, taking into account exogenous influences and pressures on development. Secondly, all too frequently, it is poverty and inequality amongst people that ultimately result in the destructive and wasteful use of the environment.

A Brief Historical Overview of Caribbean Environment and Development

The Caribbean is, of course, one of the most transformed environments in the world. As both Watts (1987) and Richardson (1992) emphasise in their respective texts, within a matter of 40 years of the initial establishment of settlers, the natural vegetation cover of small islands such as St Kitts and Barbados was completely cleared and replaced by colonial plantation agriculture, generally, tobacco and then sugar. This 'Great Clearing' organised from without let to sensational cases of soil erosion, for example, in the cases of Martinique, Guadeloupe, and the present-day Dominican Republic and Haiti, as noted by both Watts (1987) and Richardson (1992), and set the scene for present-day land degradation. Other smaller scale examples include those which characterise the Scotland District of Barbados, and the Yallahs Valley of Jamaica (McGregor and Barker, 1991).

In most Caribbean nations, agriculture remains the mainstay of the economy, with cash crops generally providing between 60 and 85 per cent of total exports and between 20-40 per cent of GDP. Nevertheless, most Caribbean countries import a very high proportion of their day-to-day food needs – frequently as high as 30 per cent (Potter, 1993a; Potter and Welch, 1994). Furthermore, the Caribbean is one of the most demographically and racially transformed environments. During nearly 400 years of slavery, an estimated 4.6 million African slaves were brought against their will into the region (Richardson, 1992).

Yet despite the fact that the economies of so many territories are so strongly based on agriculture, as shown in chapters 1 to 3, the Caribbean is highly urbanised, showing a larger urban proportion than the world as a whole. In terms of early settlement patterns, plantations dominated inland and formed the only salient unit of the settlement fabric (Rojas, 1989). All that was necessary in addition was a point of commercial, administrative and political control. As explained in chapter 2, this external imposition of mercantile and commercial control during the early colonial period witnessed strong linear-urban concentration within most Caribbean territories.

On the small islands of volcanic origin, these patterns clearly relate to physical conditions, for flat land is frequently restricted to the immediate coastal zone. However, as shown in chapter 3, they have been intensified in socio-economic terms by post-war paths to development, which have stressed enclave manufacturing, offshore banking, and recently, data processing. The region was quick to assimilate the industrialisation by invitation model. But tourism has also been causal in this respect. Whilst the region is home to 35 million inhabitants, it is also the vacation playground for over 8 million North American, European and other tourists every year, amounting to 3 per cent of the world tourist trade. As shown in chapters 2, 3 and 4, tourism in particular has given a new expression to coastal-urban development in many territories. It is also connected with cultural dependency, neo-colonialism and neo-dependency.

As in most post-colonial states, in the post Second World War period, Caribbean states have endeavoured to modernise as the royal road to catching up. Thus, Barbados in its 1979, 1983 and 1988 National Development Plans emphasised the need to modernise. Indeed, Barbados has achieved a good deal in the past thirty years, and with a GDP per capita of 9,800 at current prices in 1996, is regarded as a More Developed Country (MDC) within the Developing World. But the intention to "catch up" still remains and in the run up to the September 1994 elections, the leader of the Democratic Labour Party (DLP), David Thompson, stated publicly that by the year 2005 Barbados would be a "First World country". Although a contradiction in terms, such a statement signals clearly the intention to follow the path set by the so called developing countries. Indeed, David Thompson continued by citing in support of his assertion that United Nations data show Barbados in twentieth position in terms of GDP per capita, ahead of Ireland and Portugal. Currently, Barbados describes itself as the United National "top" developing country, by virtue of its ranking in the league table showing the United Nations Human Development Index.

However, the overriding intention of the present chapter is not to dwell on people-environment-development relations in the past, but to focus attention on their likely nature and implications in the near future, in relation to the prospects for environmental sustainability and urban sustainability in the region. It is suggested that the Caribbean currently faces two major threats which relate directly to people and the environment, and it is to these that the account now turns. Both of these stress the global dimensions of change, albeit in different contexts, the environmental and the socio-economic.

Global Environmental Change and Contemporary Caribbean Development

The first threat stems from the fact that like most Third World nations, whilst they contribute but little to greenhouse gases, Caribbean nations potentially stand to be the most affected by global warming and climate change. It has been estimated by the Intergovernmental Panel on Climatic Change (1990, 1992) that there is likely to be a regional precipitation decrease of up to 350mm per decade, accompanied by the likelihood of increased seasonality and the tendency toward fewer, but more intense rainfall events (McGregor and Potter, 1997; see also Smithson, 1993, for a more general account).

A rise in air temperatures, estimated in Wigley and Santer (1993) for the Caribbean Basin to be in the range 1.5°C to 3°C by 2060, would lead potentially to significant and wide-ranging changes in environmental conditions. For example, increases in sea surface and shallow water temperatures above a critical threshold of 30°C lead to significant increases in coral bleaching (Milliman, 1993). The immediate environmental effect of this is a reduction in the vitality of the protective effect of fringing coral reefs on Caribbean coastlines.

A significant rise in sea surface temperatures above the critical threshold of 26°C in the western Atlantic will mean a greater probability of tropical storms deepening to hurricane status. Shapiro (1982) has estimated that an increase in sea surface temperatures of 1.5°C in the region would lead to an increase in annual average hurricane frequency of about 40 per cent (from about 4 per year to between 5 and 6 per year, based on data since the early 1900s). The destructive potential of storms may also increase. Emanuel (1987) estimates that an increase in sea surface temperatures of 1.5°C would increase the potential maximum hurricane wind by about 8 per cent.

Additionally, hurricane tracks may change, due to the changing position of the Intertropical Convergence Zone, in particular with more northerly penetration. Areas such as the Leeward Islands and the Greater Antilles may become more at risk than at present (Reading and Walsh, 1995). Sea level rise associated with the expansion of ocean volumes and the melting of polar ice caps is likely to be a major threat to coastal lowlands in the Caribbean. This is especially critical for small island states, as they have high ratios of coastal lands to interior, and as much of the urban settlement and economic activity of island states in the Caribbean, including tourism, is focused on the coastal zone. The latest, and as yet unpublished, IPCC estimates of global sea level rise have been revised downward, to an estimated 5mm per annum, in a range of 3-9mm per annum.

In a review of the available evidence, Hendry (1993) estimates that regionally, sea level is rising presently at an average of 3mm per year, though considerable variation is evident throughout the Basin. He notes that the causes of this variation are not clear as yet, but are more likely related to actual sea level rise than to any underlying vertical tectonic movements. Even this reduced estimate will inevitably give rise to significant effects in the coastal zone, where urban concentrations and the tourist industry are focused throughout the Caribbean region. These are often one and the same – for example, Ocho Rios and Montego Bay in Jamaica, St Lawrence/Oistins in Barbados, Grande Anse in Grenada,

Castries-Gros Islet in St Lucia and Condado Beach, San Juan (Potter, 1995, 1996).

These physical environmental trends should be viewed in a context of rapid population increase and residential expansion in the region, due both to demographic change and the trend toward smaller household size. In St Vincent, for example, the major growth in population during the last ten years has been in the outer suburban zones of Kingstown (see Potter, 1992b). Most of these areas are mountainous, the existing settlement being on the limited coastal plain. It has been estimated that in localities such as Upper New Montrose, Camden Park and Ottley Hall, as high a proportion as 30-40 per cent of the population are squatters (Potter, 1995). Much the same can be said of River Road and Grand Anse Valley in St George's, Grenada. Thus, at the very time when there may be the chance of greater storm damage, urban residential development is occurring rapidly on less favourable land.

And this is not just happening on steeper and less stable land, it is also occurring on flatter, low-lying and marshy areas, many of which are adjacent to the sea. This is true of the Conway, Four-a-Chaud and Bananes residential communities of Castries, St Lucia. In the case of the windward coast of St Vincent, there are literally hundreds of families living on the upper parts of the beach facing the full force of Atlantic storms. Many of these residents are there because plantation owners instructed them to move to the beach (see Robertson, 1987; Potter, 1995). The passage of Hurricane Allen, for instance, on 4 August 1980, resulted in extensive damage in the Byera-Mangrove-Gorse areas of eastern St Vincent. Despite a resettlement scheme financed by the British Development Division, given the overall housing situation, a very substantial number of beach dwellers remain today (Potter, 1995). This example is more extensively considered in chapter 6.

The fact that states have provided little technical building assistance to members of the general population, in the form of information handouts, self-help manuals, peripatetic and advocacy planners, is particularly worrying in these circumstances (see Potter, 1995, for a detailed account). For example, in the wake of the passage of Hurricane Gilbert in September 1988, it was estimated that 70 per cent of all Jamaican dwellings had been damaged or destroyed, thereby obliterating most of the gains made with respect to housing in the period since independence (Clarke, 1989). As in any such storm event, the main occurrence was the lifting of flat corrugated roofs. The simple expediency of constructing traditional hip and gable-roofs would greatly reduce the incidence of such damage (Potter, 1995).

Further, the urban system is increasingly becoming unable to cope with the effects of increasing runoff, sediment and waste. Urban areas present a more or less sealed surface, particularly as sediment and urban waste choke parts of the urban drainage system. Already significant areas of Kingston, Jamaica flood during storms, and the road network increasingly becomes a temporary stream network with concomitant disruption of traffic. This problem is often at its most acute in squatter settlements, where drainage infrastructure is lacking. For example, the Four-a-Chaud area of Castries, a squatter area constructed on reclaimed land adjacent to the waterfront, regularly floods during peak rainfall events.

Four further examples serve to illustrate the varied effects of urbanisation in relation to environmental contingencies (McGregor and Potter, 1997). Firstly, at the Jamaican city scale, the Kingston Metropolitan Area and Montego Bay both illustrate the pressures of urbanisation on environmentally fragile areas, including reclaimed coastal lands liable to an increasing risk of flooding under conditions of rising sea level. Portmore, a rapidly-growing coastal zone suburb of Kingston, is particularly at risk, as some areas of urban expansion, built behind artificial barriers, are already below sea level, and as Portmore is situated in a tectonically-active area with the associated risk of earthquake and tsunami damage. Allied to potentially increased storm surge levels associated with the combination of more intense hurricane conditions and a progressively drowned protective coral reef, Portmore may be seen as a disaster waiting to happen.

Secondly, the loss of life at Weston, Barbados, in August 1995 illustrates the dangers of urban encroachment on to geomorphologically active areas within a scenario of increased risk of floodplain inundation. Following heavy rains on 2nd and 3rd August, approximately 30 houses were either partially or totally destroyed, leaving some 100 people homeless. One of the victims, the calypsonian "The Great Carew" was swept out to sea and drowned during the event. The Prime Minister, Mr Owen Arthur declared Weston a disaster area. It is believed that the gully above Weston had become blocked, alleged locally to be a result of development works on the nearby Westmorland golf course, allowing a large quantity of rain water to build up behind the obstruction. When this eventually gave way, a torrent of water and debris, including one enormous boulder, rushed down to Weston. Following the event, there was considerable discussion of the maintenance of both the west coast gullies and the hundreds of suck-wells which dot the island. In the past these were well maintained by land owners, but it now appears that many are much neglected, and are choked by vegetation and litter.

The problems caused by development in the coastal zone are eloquently illustrated by significant beach accretion and erosion patterns caused by port and power station construction in the early 1960s, and immediately to the north of Bridgetown, Barbados (Nurse *et al.*, 1995). An initial phase of strong deposition associated with port construction has been succeeded by severe erosion associated with the power station outfall discharge. Further, Nurse *et al.* (1995) note that the reef offshore from the thermal plume of the outfall is now virtually dead.

Although far from confined to damage associated with urbanisation, wetland systems throughout the Caribbean have been destroyed or prejudiced by urban development. Typically, these may be used as landfill and solid waste dumping areas, and subsequently reclaimed for building land. Bacon (1995), in a survey of Eastern Caribbean wetlands, estimates that approximately half show evidence of serious resource degradation resulting from human impact. One particular example of the nature of development is found at Graeme Hall swamp in Barbados, the island's largest inland water body, presently partly developed, and with proposals for a hotel and tourist complex.

Globalisation, Postmodernity, Environment and Caribbean Development

The second major issue relates to the process of *globalisation* and its impact on the Caribbean. Globalisation may be interpreted as a set of socio-economic and cultural processes which in certain respects lead to homogenisation, but which in others, witnesses the emergence of growing differences between world regions and localities. Globalisation thus relates to the processes of *convergence* and *divergence*, as discussed in chapter 3.

The salient argument, as presented in chapter 3, is that convergence on western norms of consumption is raising propensities to consume far more quickly than production can be increased. The net result is a rising tide of imports, a burgeoning of the informal sector, and an increasing dependency on overseas markets. A strong urban-bias to development is also the likely outcome. The role of the so-called tourist "demonstration effect" is critical in this process. The lack of indigenous principles and practices of development is noticeable throughout the Commonwealth Caribbean region. Only Grenada during the 1979-1983 Peoples' Revolutionary Government under Maurice Bishop seriously endeavoured to indigenise tourism and associated patterns of development (Brierley, 1985: Potter, 1993a; Potter and Welch, 1994). Within the dominant capitalist world system, urban areas have traditionally acted as the principal points of articulation in this complex, and frequently contradictory process of global economic, social, cultural and environmental change (see Potter, 1993c). Whilst it is true that there has been considerable 'cultural resistance' to such processes, this seems generally to have been overshadowed by the hegemony of *economic* and *political* influences in the region.

The influence of the mass media, and that of television in particular, is likely to be especially critical here. The televising by local channels of North American soap operas may well lead to a mismatch between extant lifestyles and aspirations, along with the metropolitanisation of tastes, preferences and desires. The potential scale of the trend can be ascertained from data on the use of household amenities taken from the 1990 Census for Barbados (Table 5.4). These data indicate that ownership of television and radio receivers is near universal among even low income households. Further, a surprisingly high proportion have a video recorder, even among those who live in all wood houses. The rapid spread of cable television adds a further twist to this development in a postmodern setting. Recent discussions in Trinidad witnessed that in certain areas of Port of Spain, a quarter of households are thought to have installed cable television.

It is in this light that the "sun, fun, rum" and "sand, sea and sex" images which render the Caribbean the status of an exotic beach can be interpreted, whilst the realities of highly unequal environmental and socio-economic conditions remain largely ignored. Thus, whereas indigenous views of the environment perforce stress economic values (Lowenthal, 1961; Potter, 1992c), external perceptions are being moulded to focus on a contrived view of the environment, one which carefully filters out the realities of poverty and deprivation (Dann and Potter, 1994). This fabrication can also be referred to as the "Coconut Bar" syndrome, effectively involving the peddling of paradise as a commodity, as argued in chapter 4.

Postmodernism is also associated with the hawking of heritage for the purpose of the promotion of tourism. This is well illustrated by the Barbados National Turst's Heritage Passport which offers discounts to those visiting various historic sites, including a number of plantation houses, military posts, the museum and the synagogue. The advertisement for the passport specifically invites the discerning visitor to reject the sun, sea and sand formula. But there is an associated danger here that in the interests of attracting and pleasing tourists, the real history of these sites is obfuscated or totally misrepresented. Thus, a recent advertisement for a plantation house hotel in Bequia, St Vincent and the Grenadines avers that the hotel is a place to which visitors go "not to escape the realities of life, but rather to return to them". Suffice it to say that returning to the realities of the plantation system and slavery would indeed by sobering for such tourists. As a further example, in St Lucia, a 488 acre all inclusive holiday resort has been developed on the site of the former Jalousie Plantation. This complex of cottages is set between the Pitons, one of the most scenic

Table 5.4 Availability of household appliances by house type in Barbados, 1990

Material of outer walls of dwelling	Number of occupied dwellings having household appliances in use (percentage of total occupied dwellings in brackets)											Total occupied dwellings
	Radio	Television	Video Recorder	Telephone	Refrigerator	Washing Machine	Solar Water Heater	Other Water Heating	None of these	All of these	Not stated	
Wood	25,566	21,797	8,347	13,719	21,769	1,424	286	845	1,610	273	854	30,004
(%)	(85.21)	(72.65)	(27.82)	(45.72)	(72.55)	(4.75)	(0.95)	(2.82)	(5.37)	(0.91)	(2.85)	
Wood & Concrete Block	13,971	13,968	7,282	12,026	14,210	3,098	1,256	2,204	72	1,120	201	15,090
(%)	(92.58)	(92.56)	(48.26)	(79.70)	(94.17)	(20.47)	(8.32)	(14.61)	(0.48)	(7.42)	(1.33)	
Wood & Concrete	829	843	443	714	843	197	83	148	4	76	14	907
(%)	(91.40)	(92.94)	(48.84)	(78.72)	(92.94)	(21.72)	(9.15)	(16.32)	(0.44)	(8.38)	(1.54)	
Concrete Block	23,615	23,765	14,548	21,569	24,291	11,963	9,400	4,503	96	6,961	394	25,529
(%)	(92.50)	(93.09)	(56.99)	(84.49)	(95.15)	(46.86)	(36.82)	(17.64)	(0.38)	(27.27)	(1.54)	
Stone	2,130	2,149	1,191	2,085	2,208	1,367	840	702	30	775	71	2,401
(%)	(88.71)	(89.50)	(49.60)	(86.84)	(91.96)	(56.93)	(34.99)	(29.24)	(1.25)	(32.28)	(2.96)	
Concrete	1,082	1,075	663	994	1,100	555	496	217	6	340	13	1,150
(%)	(94.09)	(93.48)	(57.65)	(86.43)	(96.65)	(48.26)	(43.13)	(18.87)	(0.52)	(29.57)	(1.13)	
Other	96	84	42	80	91	48	23	25	6	21	5	115
(%)	(83.48)	(73.04)	(36.52)	(69.57)	(79.13)	(41.74)	(20.00)	(21.74)	(5.22)	(18.26)	(4.35)	
Not Stated	7	6	4	7	7	1	1	0	0	1	8	15
(%)	(46.67)	(40.00)	(26.67)	(46.67)	(46.67)	(6.67)	(6.67)	(0.00)	(0.00)	(6.67)	(53.33)	
Barbados	67,296	63,687	51,194	64,519	18,634	12,384	12,385	8,644	1,824	9,567	1,560	75,211
(%)	(89.48)	(84.68)	(43.24)	(68.07)	(85.78)	(24.78)	(16.47)	(11.49)	(2.43)	(12.72)	(2.07)	

Source: Author's calculations; derived from Barbados Population and Housing Census (1990)

spots on the islands. The complex was initially developed by Lord Glenconner, who bought the land for £200,000 following his departure from Mustique. However, the resort, which opened in September 1992, was in the end developed with investment from several Iranian families. It was built despite a good deal of environmental outcry and opposition over the eight year planning period, not least from Sir Derek Walcott, Nobel Laureate and a national of St Lucia (Davies, 1993). The resort complex now features strongly in the offerings of a number of leading package tour operators in the United Kingdom.

The reinterpretation of history, heritage and present-day circumstances in ways which ease the consciences of tourists is, of course, all too common, as detailed in the last chapter. The important point is that such devices appear to be in the business of negating any barriers which might result in fewer tourists. The fact that such misinformation is rendering many visitors profoundly ignorant about the communities that are hosting their visits does not seem to be regarded as worthy of serious concern. But the central argument is that such machinations are blurring the boundaries between fact and fantasy in such a way as to allow states to continue to expand tourism for commercial gain, even when ecological and/or socio-cultural saturation may have been reached. In Barbados, for example, it is argued that 80 per cent of the west coast coral reefs are now dead, and most of the Caribbean Sea on the leeward side of the island has been entirely screened off by development. Insofar as Barbadian tourism seems to be becoming increasingly unsustainable – both ecologically and socially – so it is probably necessary to ponder the sustainability of that nation's entire pattern of urban and economic development in the past twenty-year period, during which tourism has increasingly invaded all parts of the society (see also Dann and Potter, 1994).

The subtlety of contemporary relationships between people and environment, fact and fantasy, and tourism and development is well exemplified by an event which occurred in late 1993 and early 1994. The Barbados Wildlife Reserve, located in St Peter, is a tourist attraction which is promoted as an "Untamed paradise ... situated in a lush mahogany forest, where uncaged animals are undisturbed in an idyllic environment". Within this promotional discourse it is noticeable once again how images of the beautiful and bountiful natural environment predominate. But in November 1993, the British Union for the Abolition of Vivisection (BUAV) published evidence of hundreds of African Green Monkeys being kept in tiny cages out of public view, awaiting export for laboratory research. BUAV mounted a boycott of Barbadian holidays during the busy season for holiday bookings, in respect of a country that had until very recently been losing tourists (Dann and Potter, 1994).

The theme of unspoilt nature also features strongly in a recent advert giving details of Outdoor Tours on the East Coast of Barbados. Prospective participants are invited to "experience the life of a 16th Century (sic) Barbadian Amerindian Carib for one wonderful day as you escape to the unspoilt areas of Barbados that are as beautiful today as they were three and a half centuries ago when the island was discovered". Leaving aside the fact that the great clearing of the island is entirely ignored, errors are compounded when it is stated that "At lunchtime the atmosphere is typically *Carib* with a three course traditional *planters lunch*". The confused environmental and historical picture is presented in an advert which depicts safari outfits, and on examination it becomes clear that the head of the figure on the extreme left of the main photograph has been altered and superimposed.

The theme of heritage being reworked in relation to the promotion of tourism is also illustrated in the Caribbean by the redevelopment of waterfront areas, such as in Curaçao, Netherlands Antilles, Bridgetown, Barbados, and St John's, Antigua (Hudson, 1989, 1998). In the case of the latter, Thomas (1991) has recently shown how, since 1980, an area of nineteenth century buildings on the waterfront in St John's has been restored and renovated. This commercially gentrified area, known as Redcliffe Quay, now includes expensive outlets such as Gucci, Benetton, The Body Shop, Little Switzerland and Colombian Emeralds. Thomas refers to the Redcliffe Quay as restored architecture for sale to the North American tourist, a process stemming directly from the fostering of an image of paradise created by the North American tourist industry, so that the "central point here is that tourism has led to the production of the built environment of St John's as a commodity for sale to foreigners" (Thomas, 1991: 482).

The expression of postmodernity clearly leads to all manner of strange paradoxes. Another example is that mentioned previously. Thus, whilst the local sustainable Caribbean vernacular architectural house form suffers from state, elite and consensual rejection (see Potter, 1991, 1995), at the same time, it is both promoted and lauded within the private and tourist sectors. This contradiction reflects the drive for modernity on the part of the state and elites, whilst the tourist sector seeks to return to the small-scale delights of the pre-modern or vernacular house form, in order to attract customers. Despite calls to use modular, moveable folk houses, state housing policy has invariably involved concrete houses – whether in the form of the barrack-like structures of central Castries, or the starter houses of Barbados. This theme is elaborated in chapter 6.

Similarly, whilst planners ignore the possible contributions of the traditional low income house, what was the multinational Marriott Hotel,

Sam Lords Castle has just as busily set up chattel houses and rum shops within its grounds, and from them serves buffet meals to its guests (Dann and Potter, 1994). Such happenings are leading to other features of postmodernism in the blurring of boundaries and the establishment of playful placenessness, as discussed in chapter 4. During June 1994, Banks (B'dos) Breweries Ltd. built a replica of a Bajan chattel house in its Wildey, Bridgetown grounds, in order to satisfy tourist's numerous requests for souvenirs. Interestingly, in the newspaper article which appeared in the *Advocate*, such houses were described as "old-time Bajan chattel houses" – or effectively as things of the past.

Thus, the yearning for the pre-modern and distinctly post-modern dream-world on the part of the tourist and the tourism industry on the one hand, contrasts starkly with the efforts of the state to modernise the economy on the other. The danger here is that whilst the postmodern tourist slips back to the past and stresses the indigenous, the host society is driven evermore to forms of development and change which reject the local and the small-scale in favour of the exogenous and large-scale. The impacts on the environment are potentially substantial indeed.

The need to restructure mass market tourism can be seen in the same light. In mature destinations like Barbados the state appears unwilling or reluctant to move from mass tourism, blaming the global economy whenever tourist numbers decline (Dann and Potter, 1994). Increasingly, the tourist industry is promoting the interests of the traveller and explorer, and eco-tourism and sustainable tourism. Perhaps there is indeed more chance that those countries which are newer to the industry, those which are more mountainous and volcanic, such as Dominica, St Vincent and Guyana, will resolve these matters in a more appropriate manner (see Britton, 1994, for example); one which provides a better balance in terms of the needs of people and of the environment. Or is this return to small-scale, ostensibly locally-sensitive tourism the next logical step in late-capitalism following the attainment of saturation in long-established resorts, and one which will be fully exploited by host states and the tourist industry? Should one be just as cynical regarding the recent emergence of all-inclusive "couples" holidays, rather than the sand, sea and sex formula, in these days of AIDS and HIV? Having saturated the potential of the mass market, is the only alternative for late capitalism to move on to specialist niche marketing for couples?

Final Conclusions

Problems of interpretation such as these confront all those who seek to examine postmodernity in the wider context. On the one hand, postmodernism involves liberation. In rejecting meta-narratives and monolothic functionalism, it can potentially focus on "Otherness" in the form of Other voices and Other worlds, and the fact that groups have the right to speak for themselves (Harvey, 1989). It is thereby associated with small-scale developments and the rejection of modernism, and is sensitive to vernacular traditions and local history. On the other hand, postmodernity can be regarded as the cultural logic of late capitalism, effectively representing the new conservatism and embracing the New Right. Thus, it focuses on the market and is strongly product-oriented. It is preoccupied with commodification, commercialisation and cheap and tacky commercial developments.

It is being posited here that it is this latter aspect of postmodernity that can most readily be discerned in the current development conundrum in the Caribbean. Here the ultimate loser may well prove to be the natural environment which will almost inevitably come to be over-exploited. Certainly the Caribbean experience briefly outlined in the present chapter would seem to bear out remarkably vividly, Harvey's (1989: 59) observation that "... postmodernism abandons all sense of historical continuity and memory, while simultaneously developing an incredible ability to plunder history and absorb whatever it finds there as some aspect of the present". The potential for accelerated social and environmental degradation in a region of almost total ecological transformation is quite literally awe-inspiring.

Finally, it is evident that the Caribbean region illustrates only too well the links between people, environments and development, both through history and at the present time. It also demonstrates how global these issues are, and have always been. The pressure to develop for short-term gain rather than long-term sustainability is likely to be exacerbated in the future as the result of both global climate change, and increasing trends of socio-cultural globalisation and postmodernity. Since many of these issues have been neglected until quite recently, attention must be placed on them over the next decade if effective paths to sustainable development are to be established in the Caribbean region as a whole, as elsewhere in Developing Societies. The alternatives appear to be new forms of colonialism and dependency, expressed through ever-increasing overseas orientation and a strong urban-based imperative.

6 Housing Conditions and the Role of the State in the Eastern Caribbean

Introduction

As mentioned in Chapter 5, the small island nations of the Eastern Caribbean - Dominica, Grenada, St Lucia and St Vincent and the Grenadines, in common with others in the Caribbean region, possess indigenous self-help housing systems that are based on expandable and moveable wooden units (Potter, 1995). The central theme of the present chapter is that such houses, built by the people for the people, represent a surprisingly neglected architectural, cultural and planning resource. As such, it provides a clear example of the local being seen as grossly inferior to the global.

Within the chapter, attention is placed on discussing housing conditions and housing quality at the outset, by recourse to available census data. Following this, attention turns to state housing policies in the region. The theme here is the lack of clear pro-active policies in the case of such states. With very few exceptions, the state has only been seen to intervene directly when the demand for land for commercial purposes has suggested the efficacy of clearing squatters and low-income groups from particular residential areas. Another instance when the state has become involved is where dwellings are threatened by the likelihood of environmental disasters such as flooding and storm damage. Examples illustrating both of these types of state intervention are provided in this chapter.

However, in almost all other circumstances, it is suggested that the state has left low-income groups basically to provide for themselves. As such, housing policies have at best amounted to a form of benign neglect. State policy it is argued, has largely failed to incorporate the positive aspects of the vernacular house form into viable housing programmes and policies. In conclusion, the chapter turns to articulation theory in order to provide a theoretical rationale for these empirical observations. It is argued that in reality low-income groups have been left to fend for themselves, as no profits are to be made from capitalist intervention. This is a far cry from the use of the indigenous vernacular house as a symbol within the tourist and commercial sectors, as discussed in Chapters 4 and 5. It is at this juncture that non-globalised and the globalised solutions to housing issues interface.

Housing Conditions in the Eastern Caribbean States

The Windward Islands are poor, ruggedly mountainous and economically dependent territories. Currently, their Gross Domestic Products per capita stand at around the US$2,000-4,000 level, whilst each of their populations amount to somewhat around 100,000. Although the tourist and construction sectors of the economy have grown quite rapidly over the last few years, agriculture - in particular bananas - remain the most important economic activity throughout the region. Industry on the other hand is generally in its infancy, accounting for less than 10 per cent of GDP among the islands. Tourism grew during the 1970s and 1980s, employing almost one quarter of the labour force of the Windwards, and currently provides between 50 and 70 per cent of the area's foreign exchange (Fraser, 1985; Potter, 1993b, 1993c).

A useful insight into housing conditions in these islands is afforded by the data provided by the census. The basic characteristics of housing are summarised in Table 6.1 for the three islands which formed the focus of a project carried out by the author, which was funded by the British Economic and Social Research Council in the early 1990s. This work concentrated on St Lucia, St Vincent and the Grenadines and Grenada. Although Dominica was visited during the project, it was not possible to put together the same data set as that collected for the other territories.

The typical household in the Commonwealth Caribbean lives in a separate dwelling unit, and flats and apartments are a comparative rarity, although this is somewhat less so in the main urban-tourist zones. This is shown to be true of all three of the Eastern Caribbean countries under consideration, with the proportion ranging from a low of 83.18 per cent in the case of St Lucia, to a high of 90.78 per cent for Grenada. As elsewhere in the Caribbean, most houses are owned by the occupants. It is vital to stress, however, that in the Census, this question referred to the legal tenure of the dwelling unit and not the land. This explains the very low number of squatters recorded in the Census.

Houses in the region which conform to the traditional vernacular form have generally been constructed entirely of wood, as shown in Figure 6.1. This pattern is undergoing a rapid transformation in most countries, with a

Table 6.1 Key indices of housing quality for Grenada, St Lucia and St Vincent and the Grenadines

Country/Area	Percentage of total households possessing the attribute:							
	Built before 1960	Constructed entirely of wood	Pit latrine toilet	Public standpipe	Separate houses	Owner occupied	Electrical lighting	Gas for cooking
Grenada	48.26	60.74	61.82	34.21	90.78	74.10	39.08	28.11
St Lucia	38.82	73.62	51.40	40.74	83.18	64.74	40.77	24.52
St Vincent	37.42	45.29	68.66	45.33	89.22	72.06	52.03	27.96
All Three	40.65	60.82	60.01	40.07	87.46	69.97	43.69	26.72
Barbados	46.78	57.31	52.22	9.96	90.70	70.19	83.02	66.44

Source: Author's research

marked swing toward building in permanent materials. But as shown by Table 6.1, even in 1980, the proportion of dwellings constructed entirely of wood remained extremely high in St Lucia, standing at nearly three-quarters of the housing stock. It is notably lower at just less than half in St Vincent. For Grenada, the proportion is just over 60 per cent, close to the average for the three nations taken as a whole.

In situations where half to three-quarters of all homes are built of wood, the age of construction of dwellings necessitates a great deal of repainting and upgrading of timbers. One way of examining this is the percentage of houses that were over twenty years old at the time of the census. Housing in both St Vincent (37.42 per cent) and St Lucia (38.82 per cent) is noticeably less aged than in the case of Grenada at 48.26 per cent. Despite progress being made in the region with regard to the supply of water, the communal public standpipe as opposed to water piped into the yard or into the dwelling itself, remains the predominant form of water supply for a very large proportion of households. This form of water supply accounts for nearly half of all connections in St Vincent (45.33 per cent). The equivalent statistics for the other two islands remain over one-third. The public standpipe is still very much part of the daily scene in the Eastern Caribbean.

A water closet type of toilet is relatively uncommon in the Commonwealth Caribbean (Singh, 1988), and pit latrine toilets are still the most frequent type of facility. The overall level of use is almost exactly 60 per cent, varying from 68.66 per cent for St Vincent to 51.40 per cent for St Lucia. An index of relative modernity such as the presence of electrical lighting shows that St Vincent fairs relatively well once again, whilst the use of gas for cooking stands at only one-quarter of households in each of the territories under consideration.

A straightforward index of housing disamenity may be constructed from the rows of Table 6.1. High percentage values on variables 1 to 4 can be regarded as signifying housing disamenity, and can be summed together. The variables 'separate houses' and 'owner occupied' measure traits that appear to vary little across the countries, and can thus be excluded from the index. The variables in the last two columns - the provision of electrical lighting and the employment of gas for cooking purposes - are effective measures of amenity. They can be subtracted from the sum of the first set of variables to give a crude general index of housing disamenity. Thus, the higher the score, the greater the inferred level of disamenity. The results suggest that St Lucia shows the poorest relative conditions, with an index score of 23.22, followed closely by Grenada with 22.97. The lowest total, and therefore, the lowest disamenity score applies to St Vincent at 19.45. To put the analysis in context, on the same

Grenada

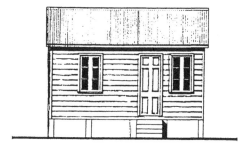

St Lucia

St Vincent and the Grenadines

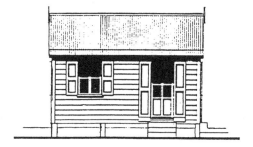

Figure 6.1   Popular houses in the Eastern Caribbean

calculation, Barbados recorded a value as low as 2.80, some ten times less than those characterising the Windward Islands. In contrast, the difference between GDP per capita for Barbados and the Windwards was approximately four to five times.

Geographically, of course, the important point is that the condition of housing disamenity is expressed geographically. The analysis of the full data set has been presented in Potter (1991a, 1991b, 1991c, 1995). One example, that of St Vincent and the Grenadines will suffice here (Figure 6.2). The incidence of public standpipe use shows a very strong north-south pattern. Levels are conspicuously high in the case of Barrouallie, Chateaubelair, Sandy Bay and Georgetown, and lower in the southern part of the island and the Grenadines. The lowest levels of public standpipe use apply to central Kingstown and the North and South Grenadines. The eastern suburban areas of Marriaqua and Calliaqua show levels of usage between 25 and 50 per cent. Turning to an index of amenity, the availability of gas for cooking purposes reaches a peak in the city centre, and in Calliaqua, and declines progressively with movement from these urban-suburban areas. Once again, the Northern Grenadines show relatively high levels of housing amenity.

The eight housing variables employed in the analysis were finally factor analysed in an effort to search for general patterns in the case of each of the three countries. The overall results of the analysis are listed in Table 6.2. In all three islands a highly general first factor was derived, which accounted for a large proportion of the variance contained in the original data set. This accounted for 69.3 per cent in the case of Grenada, 58.3 per cent for St Vincent and 57.6 per cent with regard to St Lucia.

In each instance, the first factor is a general measure of disamenity. In the case of Grenada, variables such as wood, owned houses, separate dwellings, and pit latrines load positively on the first component, while the use of gas and electricity are negatively associated with it. The same sort of pattern applies to St Vincent where public standpipe, wood, separate houses, owned and pit latrine load positively, and electricity plus gas load negatively. In respect of St Lucia, the variables owned and constructed of wood load positively once more, and the use of gas and electricity negatively.

Figure 6.3 shows the pattern of census area scores of the first factors for the three countries. In the case of Grenada, housing conditions are on average best in central St George's followed by the parishes of St George's South and St George's North in that order. St Andrew's North on the northeast coast records by far the highest disamenity score. Middle to high disamenity scores apply to St David's, St Andrew's South, St Patrick's and St John's, plus the island of Carriacou. A much lower score characterises St Mark's on the northwest leeward coast. Turning to St Lucia, the same urban-rural contrast is discernible in the scores. Progressively larger negative scores are achieved on the disamenity factor as one progresses from the rural areas of Castries quarter, to the suburbs and to the city itself. The highest disamenity scores pertain to the two southern rural areas of Choiseul and Laborie, followed someway behind by Anse-la-Raye and Canaries. For St Vincent and the Grenadines, the largest negative score on the housing disamenity factor is returned by central Kingstown, and the next highest applies to the North and South Grenadines rather than the immediate environs of the capital. This reflects the tourist- and expatriate-orientations of these islands. The next lowest amenity scores pertain to the Calliaqua area to the east of the capital city and to the suburbs of Kingstown. Housing disamenity scores reach their peak in the case of the northern peripheral census zone of Sandy Bay. On the west coast, Barrouallie exhibits a high score, as do Bridgetown, Colonaire and Georgetown on the eastern Atlantic coast. Layou and Chateaubelair show relatively low scores on the housing amenity factor.

This analysis stresses the fact that there are marked and consistent bases to the geographical variations which characterise housing conditions in the three Windward Islands under scrutiny. In particular, they serve to highlight that in overall terms, the poorest housing is to be found in the rural areas and not the principal urban zones, a point which is all too frequently either ignored or obfuscated in the realm of housing policy. Such geographical inequalities in housing circumstances reflect not only the colonial past, but also the persistence of urban bias in development since independence. However, central urban housing conditions are frequently as bad as those encountered in the principal rural zones and appear all the worse because they are found at high densities.

Housing and the State

The first point to stress is that despite the existence of such obvious housing stress, few Commonwealth Eastern Caribbean territories have an explicit national housing plan, nor a formal housing planning machinery. This is true of all three of the Windward Islands under examination here. Even Barbados, as a larger More Developed Country, only moved to this position in 1984 (Ministry of Housing and Lands, 1984; Potter, 1992a).

In a seemingly paradoxical manner, in most states in the region, the Ministry of Housing has had little or nothing to do in a direct sense with the production of houses. Such a state of affairs is partly reflected in the fact that the Ministerial function of housing is frequently combined

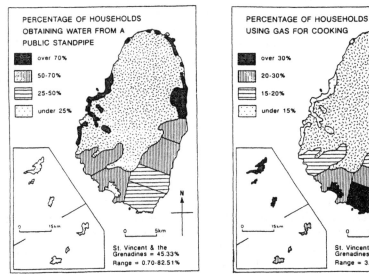

Figure 6.2   Some aspects of housing quality in St Vincent and the Grenadines

Table 6.2   **Results of the factor analysis of the housing
variables for Grenada, St Lucia and St Vincent**

GRENADA

Factor 1: 69.3% variance

Wood	+0.97
Owned	+0.94
Gas	-0.94
Electricity	-0.93
Separate houses	+0.92
Pit latrine	+0.88

Factor 2: 16.4% variance

Built before 1960	-0.95
Public standpipe	+0.58

ST LUCIA

Factor 1: 57.6% variance

Separate houses	-0.91
Owned	+0.87
Wood	+0.86
Electricity	-0.84
Gas	-0.83

Factor 2: 20.9% variance

Pit latrine	+0.78
Built before 1960	-0.55

ST VINCENT

Factor 1: 58.3% variance

Electricity	-0.90
Public standpipe	+0.89
Gas	-0.88
Wood	+0.79
Separate houses	+0.76
Owned	+0.75
Pit latrine	+0.74

Factor 2: 17.1% variance

Built before 1960	-0.94

Source: Author's research

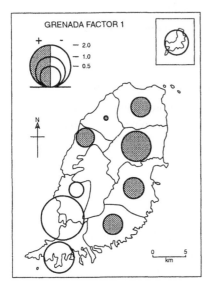

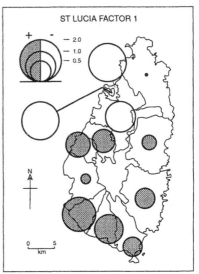

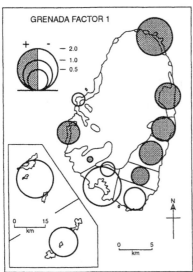

Figure 6.3　Scores of the census areas on the housing disamenity factors

together with a series of other portfolios. For example, in 1991, the Ministry of Housing in St Vincent and the Grenadines was linked with Local Government and Community Services, and in discussions, the then Permanent Secretary stressed that there was no officer who specifically dealt with housing issues. In a similar vein, in St Lucia, the Ministry of Housing forms part of a wider Ministry of Health, Housing and Labour, and has undertaken no housing projects whatsoever. Phillip (1988: 59) comments that the "Ministry is therefore the least effective and active in the field of housing".

This situation is reflected in the miniscule proportion of the housing stocks of these three territories which is provided by the State, as summarised in Figure 6.4. The government rented housing stock is largest in St Lucia at 1.50 per cent of total households. However, the proportions are as low as 0.18 and 0.36 per cent in the case of Grenada and St Vincent and the Grenadines. This may be compared with a figure of 4.50 per cent in Barbados (Figure 6.4). In the case of St Lucia, for instance, the state's involvement in housing dates from the fire which destroyed Central Castries in 1948. Between 1949 and 1959, the state built 357 apartment blocks in Central Castries. The units vary from one to four bedrooms, and in 1990, fetched rents from $70 to $200 per month. In the final analysis, these units were deemed totally inappropriate, whereupon the state has almost entirely withdrawn from the housing market.

In all three nations under examination, the function of providing and overseeing housing is vested in a technically-oriented national housing agency. In St Vincent and the Grenadines this is referred to as the *Housing and Land Development Corporation.* In the case of St Lucia it is the *Housing and Urban Development Corporation,* while in Grenada it is styled the *National Housing Authority.* These units are expected to be self-financing, working on the basis of cost recovery and replicability, by means of purchasing land for the next scheme from the money recouped from the previous one. It is not expected that they will receive funding from the state, nor that they will provide lower income groups with subsidised housing.

The overall result is that very few houses have been built in the public sector, and a minimal number of improvement programmes have been carried out. Further, the few houses that have been constructed by these respective Corporations have turned out to be well beyond the means of those who are most in need. For example, in St Lucia, the Housing and Urban Development Corporation was originally established in 1971, but was totally dormant from the early 1980s to 1989, when it was reconstituted. During its 'active' period, however, it was involved in only one scheme which was specifically geared toward the lower end of the housing market. This was the construction of 110 timber houses at Independence City in the Entrepot area of Castries. The units, which were built in 1975, measured about 39 m^2. Although they were ostensibly designed for low-income earners, the fact that they were sold at $EC 11,000 meant that a significant proportion of the houses were eventually purchased by middle-income earners. In another, somewhat earlier scheme, undertaken at Sans Souci in 1971, masonry units of some 93 m^2 were constructed. These were priced between $12,000-$14,000, and as a direct result, were also purchased by middle-income earners.

In contrast to this manifestly poor performance, 1980 saw the establishment of the *Housing Rehabilitation Project* (HRP) in St Lucia, following the destruction caused by Hurricane Allen in August of that year. The scheme was, in fact, physically located within the Housing and Urban Development Corporation, despite the longstanding quiescence of the latter. The aim of the HRP was to provide direct assistance to the 1,940 families whose homes were devastated by the storm. As Louis (1986) has effectively demonstrated, the salient point of the project was that many of the housing units provided were factory pre-fabricated and were transported to the site of construction. Thus, he argues that with proper management, stringent financial control and an effective loan recovery operation, the prefabrication of houses for the low-income could well be a feasible venture in the region. When informal interviews were being conducted by the present author in late 1990, the General Manager of the Housing and Urban Development Corporation said that they were looking at the possibility of introducing a two room, one bedroomed starter house, designed for expansion by the occupant over time. It was envisaged at that time that such units would sell at $30,000 or thereabouts.

In a remarkably similar manner, it has been argued that the public sector *Housing and Land Development Corporation* (HLDC) in St Vincent has failed to provide effective low-income shelter. The Corporation was established in June 1976 by an Act of Parliament. Ishamel (1989) in a report made to the United Nations Center for Human Settlements (UNCHS) puts the failure down to a mixture of internal mismanagement and the lack of clearly designed housing policies in the first place. Although charged with remaining economically viable, from its inception, the Corporation always provided some subsidies. This was true in the case of the sale of building materials by HLDC. The criteria for eligibility for subsidised building materials were a maximum income of $15,000 per annum, total assets of $40,000, whilst an upper limit of $60,000 was placed on the house price. However, it was well-known that relatively wealthy people were purchasing building materials from HLDC at discounted rates. By the financial year 1989/90, the Corporation had run up a $1.9 million

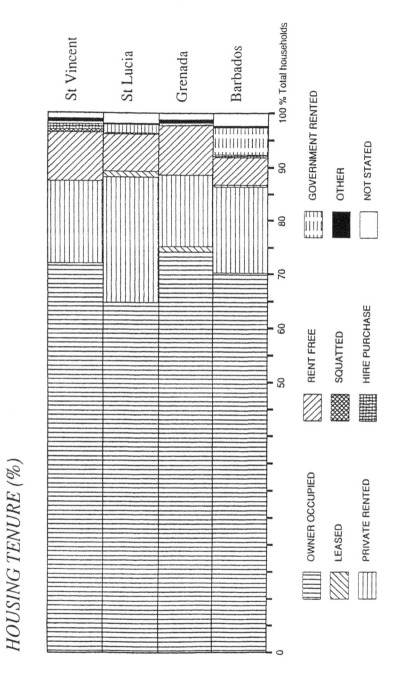

Figure 6.4 Housing tenure in the Eastern Caribbean

overdraft, and the public image of the Corporation had become severely tarnished. Indeed, Ishmael (1989) recommended that HLDC should be dissolved legislatively, and a new housing finance institution be established in its place, together with its own integral technical service agency.

Over the past few years, however, there have been signs of a changing awareness in the case of St Vincent and the Grenadines. In September 1988, the then Minister of Housing outlined the Government of St Vincent and the Grenadines' intended approach to housing. The enlightened point here was that it was announced that future policy was to focus mainly on the provision of site and services schemes and the upgrading of existing urban low-income housing areas. Since then, HLDC has been endeavouring to upgrade two areas per year, with each of these consisting of 25-30 houses. This is a small total, but a step in the right direction. In areas such as Upper New Montrose and the Malla Valley, footpaths have been improved and two standpipes provided. It is envisaged that other upgrading efforts will focus on areas such as Camden Park, Routcher Bay, Coconut Range, Gibson Corner, Fair Hall and Largo Heights.

In discussions with the author in 1989, the then General Manager of the National Housing Authority (NHA) of Grenada confirmed that there was no national housing plan. Further, given that Government to that point had thought exclusively in terms of producing entire houses, there was seen to be no need to apply to external agencies such as the World Bank or USAID for housing funds. It was further reported that only at that stage were the Government and NHA thinking of establishing a small site-and-service scheme at Corinth. In fact, between 1975-1989, official statistics show that national housing schemes saw only 327 houses built (Grenada Planning Office, no date), and most of these were affordable only to middle income families (Wirt, 1987).

The essential message of this section is that the Eastern Caribbean countries under study do not currently possess overt national housing plans in any shape or form. Their respective Ministries of Housing have been dormant, and the responsibility for improving the housing lot of those of low-income has been placed at the feet of technically-oriented national housing corporations. These have been expected to run on the basis of cost recovery, with minimal subsidies being transferred to the poor. As a result, most of the houses constructed by these divisions have been way beyond the means of those on low incomes. In the case of St Vincent and the Grenadines, there has been a recent step towards a more pro-active policy, with the espousal of the need for squatter upgrading programmes on the steep hillsides around the capital Kingstown. This has followed a period during which the national housing agency was far from successful. However, in both St Lucia and Grenada, there are no upgrading or rehabilitation schemes for the existing housing stocks.

The Question of Housing Finance

In general terms, the housing finance opportunities available in these Caribbean territories, like those in many Third World nations, mean that the urban and rural poor are largely, if not entirely, excluded from the opportunity of obtaining a mortgage given the price of housing and land in the formal sector.

For example, in the mid 1980s, the cheapest unit constructed in the formal sector in St Lucia cost $33,000 (Phillip, 1988). At a mortgage interest rate of 12 per cent per annum, over a repayment period of fifteen years, the general terms prevailing at that time, such a house was not affordable to the lowest 60 per cent of income earners. At about the same juncture, Central Planning Unit data show that approximately 300 units were being produced per year by the formal building sector. The reported average cost of these units was $86,191, which could only be afforded by the top ten per cent of income earners. Hence, although funds for lending are available through the St Lucia Development Bank, the St Lucia Mortgage Finance Corporation, the National Insurance Scheme and the St Lucia Co-operative League (comprised of a group of credit unions), these are all effectively out of the reach of low-income house searchers (see Louis, 1986: 64-68 for further details).

In St Vincent and the Grenadines, there are approximately twenty institutions, mainly banks and credit unions, that are involved in dispensing mortgages. Since 1988, the main source of mortgage finance has been the Government-owned National Commercial Bank (NCB). The funds derive from a loan of $4.05 millions from the Caribbean Development Bank, together with a matched pool of funds from the National Insurance Scheme, making $10.7 million available in all. NCB loans can be used to cover 90 per cent of the market value of land and new construction. However, with a ten per cent interest rate over 20 years, the bulk of funds have served the needs of only the top 25 per cent of Vincentian income earners. Ishmael (1989) notes that approximately 75 per cent of the population earn under $10,000 per annum, whilst the bottom 20 per cent earn less than $3,000 per year. Even if such groups could afford a mortgage, the maximum purchase they could manage would be in the $18,000-20,000 range. The cheapest unit available in the formal housing market at that time cost $35,000, close to the average cost in the case of St Lucia.

Thus, low income groups in these societies are unable to service their shelter needs within the formal housing, finance and credit sectors. At the same time, Government is not involved in providing housing or indeed housing subsidies. Hence, the bottom 60-75 per cent of income earners have no option but to provide their own housing by whatever means they can. In short, as is replicated in so many contexts, those groups who are most in need of housing are not reached by the formal sector, nor the state.

Urban Squatter Settlements, Rent Yards and Shanty Towns

The account thus far has detailed how poor groups within the societies under study are marginalised and excluded from the formal housing process. Those with low incomes have no option but to provide their own homes by informal means. As a result, squatting has become more and more common around the capitals of these territories, especially on the steep hills which surround Kingstown, Castries and St George's. Some authorities talk in terms of 30 per cent of the households in such areas being squatters. Squatting has become particularly prevalent on land which is marginal for other purposes, for example, on steep slopes and poorly drained land, as well as on Crown or public lands. A recent analysis of the preliminary results of the 1991 Census for St Vincent and the Grenadines shows the rapid growth of population in the Kingstown suburbs and Calliaqua, with growth rates well over 16 per cent between 1980 and 1991 (Potter, 1993b). In contrast, Kingstown proper actually lost population during this period (Figure 6.5).

There are two major areas where people have been squatting for some time in Castries, St Lucia. These are both shown in Figure 6.6. The first is the Conway area located just to the north of the Central Business District (CBD), adjacent to the harbour. The second is the Four à Chaud area which is again located adjacent to the harbour, this time directly to the west of the CBD. The Four à Chaud area was originally settled in 1972, and today houses approximately 428 dwellings with a population in excess of 2,000. The area was originally swampy and residents stress that they were encouraged to settle there by members of the then government, albeit on a temporary basis. Phillip (1988) in a study of the area stresses that many of the inhabitants felt that they owned the plots on which they reside, as a direct result of the fact that they filled in the lots themselves, having paid truck drivers to deliver soil to their plots in an effort to make them suitable for building. However, the origins of the area are shown by the fact that it is still very prone to waterlogging and periodic flooding during the wet season. Drainage in the area is very poor, consisting of a few open drains. There is only one public facility comprising a shower and toilets and a single public standpipe which is reported as serving the 153 households that do not have water piped into the house or yard. As high a proportion as 70 per cent of the population use the public facility as the means of sewage disposal, 22 per cent pit latrines and 8 per cent 'other' methods (Phillip, 1988). The land is ultimately zoned for warehousing activity as a part of the port zone, and Phillip (1988) reports widespread concern among the residents about the possibility of relocation.

The other main squatter area of Castries, known as Conway (Figure 6.6), is built on land which was originally owned privately by a single family and which was being developed as an urban rent yard. The area grew rapidly during the 1940s and 1950s as a result of the flourishing trade which occurred in Castries harbour during the period. Stevedores, in particular, needed to be near the harbour and so were prepared to pay the low rents charged and build their own homes on an informal and largely *ad hoc* basis (Phillip, 1988). Eventually, the owners of the land felt that they could not keep abreast of the rent yard-type of development which made rent collection difficult. They, therefore, sold the land to the Government in the early 1960s, whereupon the area effectively became a squatter settlement as the state did not bother to collect rents for the spots. By this stage the area consisted of well over 300 houses.

But the Conway area affords an interesting lesson concerning the role of the state in housing matters in the region. Generally, as elsewhere, the state has not intervened and squatters have been left to get on with the job of building for themselves. But the state has recently become involved in the case of the Conway area. As long ago as the 1970s, the area was scheduled for commercial redevelopment. The plan envisaged a central Piazza, with commercial offices being built along the waterfront, facing the John Compton Highway. Nothing happened, however, for approximately 20 years. Then in the middle of the 1980s, a new duty-free tourist facility was developed across the harbour at Pointe Seraphine. In 1986, after the opening of the duty-free facility, the Government announced its plans to clear part of the Conway area - the first intervention of its kind in St Lucia - and build commercial and governmental offices. The western part of the squatter zone was cleared with considerable haste between April and July 1986, with a total of 162 houses being relocated in three phases. These former residents of Conway were moved to Ciceron, some three kilometres to the south-west of the CBD. The Housing and Urban Development Corporation had land here, and the inhabitants of Conway were given the incentive of $1,000 either as a direct deduction from the cost of the land, or in the form of materials for renovating the unit. The Government provided free transport for the dismantled houses to be carried to their new sites. Two meetings were held with members of the area, but both took place

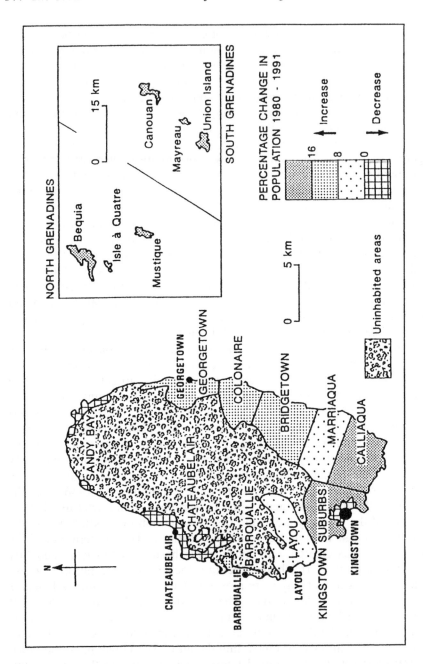

Figure 6.5 Population change 1980-1991 in St Vincent and the Grenadines

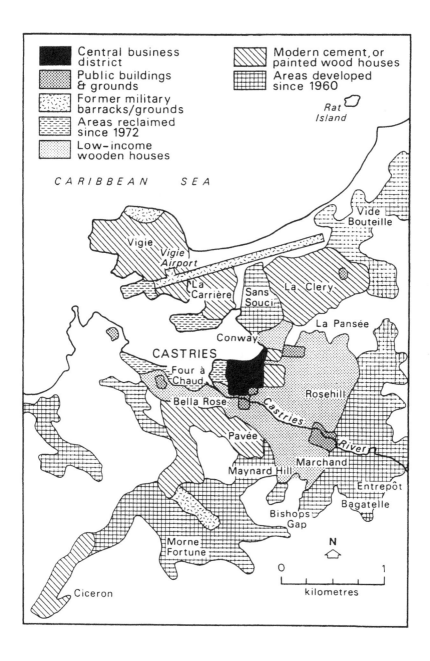

Figure 6.6   Land use patterns in Castries, St Lucia

after the relocation process had itself begun.

It seems clear that the process of state involvement in housing after a period of extended quiescence reflects the interests of capital. The moves were made on economic grounds, in order to release prime, valuable land in the city centre. It is hard not to ponder whether it also witnessed the desire to remove what might be considered as an eyesore in the vicinity of a large new tourist-oriented development.

A noticeable recent development in Castries has, however, been the evolution of new squatter settlements close to the urban core, and it is hard to envisage that these two sets of circumstances are not very closely linked. The first has occurred where squatters have started to build houses in and around the small cemetery which is located just to the east of the Four à Chaud area. The second area is one farther along the harbour from Four à Chaud, at Bananes Point, which is in the process of being developed as a self-help fishing community.

Essentially similar examples could be presented in respect of both St Vincent and Grenada. Suffice it to say here that squatting has become an all too frequent occurrence in, and around, Kingstown, St Vincent. Ishmael (1988) reports that most squatting occurs on Crown Lands and is strongly based on the apparent lack of past reaction on the part of the authorities. Some local experts argue that this explains why the housing stock of St Vincent is found to be better than those in other states. Central Kingston, in particular, is surrounded on the east by rapidly growing squatter communities which are extending up the mountain slopes. This is, of course, exacerbating problems of deforestation, soil erosion and increasing agricultural resource tensions. There are well-established squatter settlements in Upper New Montrose and Malla Valley, where densities are high due to infilling. A recent development in St Vincent appears to parallel the one elaborated above in relation to Conway, St Lucia. There are a number of squatter areas on the slopes which overlook the south-west running valley which runs down to a very scenic beach at Ottley Hall to the north-west of the capital, Kingstown. For some time it has been intended to develop this area as a marina and resort, and there has been talk of clearing the squatters from the area as a part of this. Planning permission has been granted and a $135 million marina and luxury cruise ship port will be developed over the next few years.

Evidence from the region confirms that housing is a strongly political issue, with squatting being actively encouraged and supported by politicians in the periods immediately prior to elections. At the same time, existing shanty and squatter areas are likely to see the provision of basic services during such periods. In a survey carried out by Ishmael (1988), 78 per cent of those interviewed in two areas claimed to have been authorised to squat either by politicians or HLDC officials. After elections, the settlements continue to be neglected. In this way, self-help housing may be used as a political tool in return for electoral support. At the same time squatters have, with the few exceptions outlined above, rarely been disturbed by the authorities. Thus, in St Vincent, in particular, walled houses are quite common, and in 1981, for instance, formed a larger proportion of the total housing stock in this impoverished country (65 per cent) than they did in Barbados (45 per cent). It is worth quoting a regional writer here, Louis (1986: 275) that "one would hate to believe that squatter settlements are a deliberate method of satisfying housing needs without the massive investments that would have been expected, had it been a controlled development".

Another instance where the state may become involved in housing is where environmental factors threaten the safety of settlements. An instance of this was provided earlier in relation to the rehabilitation of housing after the passage of Hurricane Allen in St Lucia. A second example is provided here, and it is one which also serves to exemplify the marked lack of emphasis which is accorded low-income housing in general in the region. On mainland St Vincent, a large number of self-help houses are to be found located on the upper parts of the beach on the windward side of the island. A large constellation of such houses is to be found in the Mangrove, Gorse and Byera areas. Similar communities are to be found just to the south of Georgetown.

The precise location of such dwellings along the Mangrove, Gorse and Byera section of the east coast is shown in Figure 6.7, which has been adapted from Robertson (1987). It is estimated that the beach in this area is home to about 116 households. The history of these dwellings can be traced to Hadley Brothers Enterprise Limited, a plantation growing coconuts and bananas. In the early 1960s, all households on the estate were ordered to relocate onto the coastal beach flats. At first, such 'lands' were occupied free, although HLDC files show that subsequently the occupants were asked to pay 'peppercorn' rents amounting to $12 annually. In 1973, in a further development, the occupants were asked to purchase the spots at 10 cents per square foot in the case of labourers, and 20 cents per square foot for non-labourers.

However, with the passage of Hurricane Allen on 4th August 1980, extensive flooding caused damage to housing, possessions and livestock along the Gorse-Mangrove-Byera area. Following these events, the coastal beach was regarded as a danger zone for housing. Accordingly, a proposal for a scheme to relocate the beach dwellers was put before the British Development Division (BDD) on 8th December 1980. The total funding required was estimated at EC$ 448,271, with approximately three-quarters

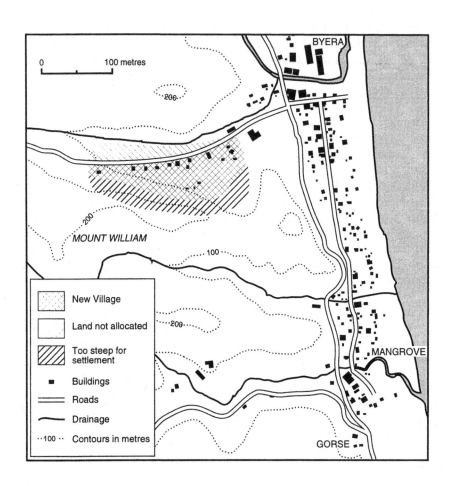

Figure 6.7 The Mangrove-Gorse-Byera area and resettlement village

coming from BDD. The land for a resettlement village was made available by the Mt William Estate. The precise location of the new inland village is shown in Figure 6.7.

Progress with the scheme was very slow at first, but by 1985/86, some 60 per cent of the original households residing on the beach had been relocated to the new village. However, as might be anticipated in the situation of a housing shortage, there remain to this day, a substantial number of beach dwellers. Clearly, the children of some older housing units remained in the beach houses when their parents moved, whilst in other instances, new residents probably moved into the zone. Thus, some thirteen or so years after the original incident occurred, the threat of severe flooding and storm loss remains, testifying once again to the "afterthought" nature of housing in the Eastern Caribbean.

Conclusions

These facts add up to a form of non-policy with regard to low-income housing. The onus of producing housing has remained firmly and squarely with the poor themselves. Most strangely of all, the Governments of Caribbean territories have rarely seen fit to place the local vernacular cottage at the centre of systems of state aided self-help. In this sense, they have not planned *with the local people* and their distinctive socio-cultural innovations. This may be situated in the debate between liberals such as John Turner who argue that the state should confine its attention to helping the poor to help themselves (see Turner, 1967, 1968, 1976, 1982) and Burgess (1982, 1992), who from the radical stance, argues that taken to extreme, self-help principles allow the state to abdicate its responsibility within the housing arena (see also Conway, 1982; Potter and Conway, 1997).

Some would argue that this is because there are just too few votes to be had from providing appropriate aided self-help housing in the region. More votes are to be obtained from legitimising or otherwise spontaneous self-help houses. In addition, the argument runs that symbolic schemes, which involve providing a few houses at a high technical specification have far more visual and political impact. This also helps to explain why housing conditions are found to be so much poorer in the rural areas of these territories, a feature pinpointed earlier in this chapter. Further, leaving the poor to fend for themselves helps to maintain a pool of labour, affords the context for powerful patron-client relationships and formidable patronage systems, as well as serving to maintain relatively low wage levels. This applies in both the urban and rural areas, of course.

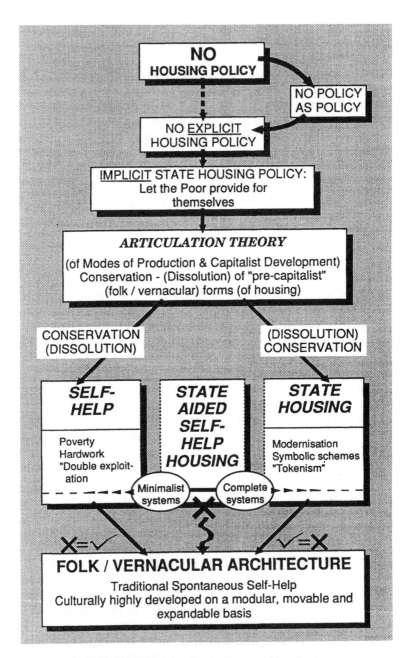

Figure 6.8 Housing policy in the Eastern Caribbean: a theoretical framework

The implications of these observations for the development of a theoretical framework for the consideration of eastern Caribbean housing are summarised in Figure 6.8, which has been drawn up by the present author. The failure to produce explicit housing policy statements can be seen as an implicit state policy of allowing the poor to fend for themselves. Thus, no policy must be seen as a clear policy (upper part of Figure 6.8). This can be interpreted in the light of ideas emanating from *articulation theory*, concerning modes of production under capitalist development (see also McGee, 1979; Wolpe, 1980; Burgess, 1990). In simple terms the argument goes that capitalism *conserves* traditional precapitalist forms that work in its favour, but specifically seeks to *dissolve* those which do not.

Thus, the stance of governments may be interpreted as one of primarily seeking to conserve the "pre-capitalist" or "informal" housing system in its most basic form (Figure 6.8). This is denoted on the left side of the diagram. This process involves the fostering of strict self-help, associated with minimalist systems of state assistance. It is linked to poverty, hardwork and exploitation at both work and at home. On the right side of the diagram, the brief flirtation with complete state housing systems, consisting of too high a specification, can be seen to be leading to the ultimate *conservation* of the folk/vernacular architectural form at its most *basic* level, although at first this seems to be leading to dissolution. Such a situation involves modernisation, symbolic schemes and tokenism. However, the intermediate position of improving such housing by means of appropriately financed state aided self-help housing has largely been ignored, as shown in the centre section of the figure.

Such a policy stance would involve the provision of small, low-interest loans and technical assistance on site. It would almost certainly entail producing core units associated with wooden houses built along traditional lines. It would witness the enhancement of the local vernacular architectural system, not its maintenance at a most basic and exploitative level. The Caribbean vernacular house - built by the people for the people - is a surprisingly neglected cultural artifact in the Caribbean region. This is exemplified by the general lack of scholarly attention that has been paid to such housing. It is also reflected in the fact that in the realm of state housing policy and provision, the efforts of low-income groups to house themselves have been almost totally neglected as a socio-cultural resource of major importance (Potter, 1989a, 1992a).

For far too long, the policy potential of the Caribbean popular dwelling has remained unrecognised by politicians and development planners alike. The work carried out as part of this project suggests that, only when the appropriate, basic needs, and self-help characteristics of such people-built houses are fully recognised, will genuine progress be made in improving low income housing in the region. Just as the grand Georgian house may be interpreted as the outcome of imposed colonial power, so the vernacular architectural form can be construed as a form of socio-cultural resistance to the inequities of that earlier colonial domination.

7 Environmental Impacts of Urban Development and the Urban Informal Sector in the Caribbean

SALLY LLOYD-EVANS and ROBERT B. POTTER

Introduction

As implied in chapter 5, throughout centuries of change and development in the small island states of the Caribbean, human interaction with the physical environment has remained highly visible. Indeed, one of the principal themes of this volume is that the Caribbean is one of the most highly transformed and urbanised regions in the world (Potter, 1989a, 1993c). Thus, rapid urbanisation and recent economic change have substantially altered both the natural and social environments. As Demas (1965) and Worrell (1987) have both observed, the contemporary Caribbean is characterised by small territorial size, undiversified economies and inherited economic dependency. Although most of the islands are politically independent, they are still subject, in varying degrees, to external economic and military domination and control.

The region has been subjected to centuries of imposed external power. This domination has continued through slavery, to the importation of western models of development involving industrialisation and modernisation (Lewis, 1955), and more recently, to economic restructuring agreements with the International Monetary Fund (IMF) (Thomas, 1988; Lundy, 1999). Unfortunately, political independence has not ended the external influences and pressures placed on Caribbean environments. Recently, attention has focused on environmental degradation in the region, as a direct result of colonial policy, and recent activities such as tourism, urbanisation and industrial development (Thomas, 1988; Potter, 1989a; Deer *et al.* 1990; McAffe, 1991; Richardson, 1992). Thomas (1988) notes that the term *crisis* is the most widely used to describe prevailing social, economic and environmental conditions in the Caribbean. The crisis can be attributed to growing poverty, economic collapse, the disintegration of domestic food production, inappropriate industrial policies and rapid urbanisation.

This chapter explores the nature of environmental degradation in the region, with specific reference to recent industrial, commercial and urban change. Particular attention is paid to the growth of the informal sector of the economy which is presenting the region with a new set of environmental concerns. Lampart (1993) states that the issues which are in need of immediate attention are pressures on the coastal zone, the nature of the informal economy, practices in subsistence agriculture and the impact of tourism.

Economic Change and Environmental Degradation Over Time

Environmental degradation, as a direct result of economic change, is not a recent event in the Caribbean. Columbus' voyage of 'discovery' represented a spatial expansion of political and economic control (Wolf, 1982) which had great influence on the natural environment. The Caribbean has been part of the international system of trade for four centuries, a process which is best exemplified through the region's integration within the international division of labour. The effects of slavery, emigration and foreign industrial investment are widespread. The Caribbean's external focus has been a result of its historical introduction to trade since the 1400s, the environmental effects of which have been far reaching.

The islands were heavily forested and the vegetation diverse when Columbus first arrived in the region (Richardson, 1992). The indigenous peoples who inhabited the islands prior to the Spanish conquest, the Carib and Arawak Indians, disappeared rapidly after conquest due to a combination of enslavement and disease (Watts, 1987). The decline of the indigenous population marked the first stage of environmental devastation in the region.

To the Spanish, the Greater and Lesser Antilles provided land for imported agricultural techniques, and new breeding grounds for animal husbandry. As Spanish agricultural staples did not fair well in the region due to the climate, Columbus brought in sugar cane, the crop which has become synonymous with the Caribbean. Rapid deforestation of the land for planting cane was undertaken on a large-scale, using imported African slaves. The removal of vast amounts of forest also produced lumber for firewood and construction, a move which represented a further step in the massive environmental change that was to transform the region. Abandoned villages, widespread sugar cane cultivation and savanna lands turned over to cattle grazing, led to rapid soil erosion and leaching. Large areas of forest were cleared rapidly, and by the mid 1600s, there had been extensive clearance in the Lesser Antilles.

The subsequent conquest of many of the islands by the British, Dutch and French only enhanced the mismanagement of the environment, particularly in relation to the plantation economy. Colonisation was a gradual process, which started with the cultivation of crops such as tobacco, cotton and cocoa, and later sugar cane. Many islands were turned into monocrop economies, as the plantation competed heavily with local food crops and forest. The islands became absorbed in the expanding European commodity exchange, over which they had no control. From the 1600s, the growing demand for sugar cane in Europe gave little regard for the widespread social and environmental devastation which was taking place as a result of the plantation system.

The plantation combined factory and field at an early date, and effectively represented 'industry' prior to the rise of industrial production in Europe (Mintz, 1985), an argument which links directly to the evolution of plantopolis (chapter 2). The Caribbean plantation unit revolved around the production of sugar cane through the utilisation of imported slave labour under European management. Plantation agriculture in the region has been well-documented by a number of academics, including Beckford (1972) and Sheridan (1973), but the ecological impact of the colonial sugar cane plantation has received less attention than its social and demographic dimensions (Richardson, 1992). The cultivation of sugar cane, combined with widespread forest clearance, represented an ecological discontinuity with the past. The most spectacular documented case of environmental devastation in Caribbean history was the burning of the forest and scrub of the entire island of St. Croix by the French "to make the islands more healthy" for the European settlers (Dirks, 1987: 16).

The immense influence of the plantation as the dominant unit of production is displayed throughout the Eastern Caribbean in terms of settlement pattern and land use, as discussed in chapter 2. In a similar manner, land use transformation followed the demand of the plantations. Plantation requirements stretched beyond the demand for cleared land, as large amounts of timber were also needed for construction and fuel (Sheridan, 1973). The islands were not just subject to environmental degradation, they were also biologically transformed. There was the introduction of new plants and animals, many of which added to the destruction of indigenous vegetation. As larger areas of land were cleared, European settlers favoured a move to the coast, as in Guyana where sea walls were constructed on mangrove swamps and mud flats. The situation changed little after the emancipation of slaves in the 1830s, when there was the introduction of new cane factories and railroads to aid economic production. In this way, the environment has constantly been degraded by uncontrolled economic practices, most of which are still subject to external control.

The colonial legacy is one of sugar cane domination, supplemented by the export of other tropical staples. During slavery, the production of local subsistence crops had been limited, a problem which has been carried into the present agricultural system. Many Caribbean nations still have to import a large proportion of their fruit, vegetables and staples. The importance of the physical degradation of the environment, which resulted from colonial control, cannot be over-emphasized. Food deficits are still a problem throughout the region, and the inability of households to expand incomes through subsistence crops, due to infertile soil, is a major problem in times of hardship. The colonial powers had complete domination of the total environment of the Caribbean in a way that was not seen elsewhere. The land now inherited by the people of the Caribbean has been used to gain profit for centuries, in a similar process to the utilisation of their labour by foreign powers (see Besson, 1987; Potter, 1992c). Caribbean degradation is as much a legacy of the past as it a result of the twentieth century.

Today, external economic control, in the form of foreign industrial investment, tourism and structural adjustment, has introduced a new set of environmental problems. The region's agricultural and mineral exports are still dependent on external markets, and tourism is crucial to the economic survival of many countries. In this respect, the economic benefits from the Caribbean environment are still being accrued mainly outside the region (Lowenthal, 1987), whilst the disbenefits are still affecting the islands themselves. The implementation of structural adjustment packages by the International Monetary Fund (IMF) has further restricted public spending across the region, and is leading to a marked increase in unemployment, poverty, and further degradation of the natural and social environments (Lundy, 1999).

Current Environmental Pressures in the Caribbean

It has been argued by McAffee (1991) that the processes which keep wealth flowing from the South to the North violate the natural environment and undermine peoples' ability to earn a livelihood. In the Caribbean, the growing separation between those who control the process of production in industry and agriculture from those who provide the labour power, results in the failure to implement sustainable development. Similarly, escalating debt often renders Caribbean nations incapable of giving the environment due consideration. Understandably, the struggle for daily survival often takes precedence over more environmentally sensitive, and costlier, practices in employment, housing, industry and agriculture. As in previous centuries, environmental degradation caused by inappropriate economic and urban activity, is clearly visible in the Caribbean today. Table 7.1 highlights a range of current social and economic trends in the region which is giving rise to environmental concern.

Firstly, the importance of export agriculture over domestic production is resulting in the use of pesticides and fertilisers to enhance exports. The inability of many countries to feed their population is further exacerbated by a declining ratio of arable land per head of population, which is a direct correlate of rapid urbanisation. Secondly, there is the inappropriate use of forest and water resources for energy consumption, tourism and industry. Thirdly, rapid growth of the tourist industry has led to a construction boom which results in the depletion of sand and other mineral deposits from valuable beach resources. Furthermore, unchecked urban construction together with the clearing of swamps and forest for the tourist industry, damage the region's natural beauty and its capacity to produce food for its growing population. Deforestation, with its multiple ecological consequences, long a problem in Haiti and the Dominican Republic, is now spreading to other islands due to increased demand for energy and timber. The result is seen in the tonnes of topsoil that are either washed into the sea every year, or which cause accidents in haphazard housing developments. Finally, the dumping of foreign waste, periodic industrial accidents, and the abuse of pesticides in an effort to grow more crops, all threaten the long-term sustainability of the natural environment.

In the last century, the Caribbean suffered from the adverse effects of agricultural degradation due to the plantation economy and slavery. Progression into the twentieth century has seen pressure for many islands, including Barbados, Jamaica and Trinidad, to modernise through industrialisation, a transition that has brought with it problems of waste disposal and pollution control. The Caribbean Basin Initiative (CBI) attempts to solve the region's economic difficulties by encouraging an increase in manufacturing exports. As explained in chapter 3, Caribbean governments are encouraging foreign investment, in the form of manufacturing plants and data-processing companies, by means of incentives such as low wages and the curtailment of environmental policies (Safa, 1990; Pearson, 1993).

Whilst the problems of environmental degradation have increased in the 1990s, Caribbean governments have seen their ability to cope with the consequences being eroded. The region is much poorer today than it was a decade ago due to the effects of the adverse balance of payments, falling

Table 7.1   Current environmental problems affecting the Caribbean region

Environmental Issue	Causal Factors
Declining ratios of arable land per capita	Urban Growth
Unequal balance between agricultural production/consumption	Decline in domestic crops
Soil erosion, infertile land, cleared forests	Past land use
Forest usage for energy and manufacturing	Tourism, Informal Industry, Foreign Industry
Limited water availability	Tourism, Industrial Development
Over-fishing and hunting of rare species	Export Industry, Informal Sector
Removal of sand from beaches	Urban Construction, Tourist Developments
Pollution (lack of control and limited infrastructure)	Over-population (tourist), Irregular Housing, Industrial Development, Informal Sector
Degradation of marine environments	Tourism, Industrial Development
Degraded urban environments	Uncontrolled Urbanisation, Inefficient Resources, Poverty, Lack of basic services

real wages, and debt and austerity packages which have led to escalating unemployment and poverty. These pressures have served to destabilise the region, whilst their economies are still heavily dependent on the international arena. In this way yet again the global is dominating the local. In their desperation to obtain hard currency to service interest repayments, governments are also inviting investors to fell more forests and replace more food crops with export crops (McAfee, 1991). Furthermore, the insular and small-scale nature of these island states makes them ill-equipped to cope with the natural disasters which plague the region. The world is fully aware of the devastating effects of Hurricane's Hugo and Gilbert which brought entire countries to a state of emergency.

Socially, increased poverty and the decline of social service provision has lead to the growth of urban blight and increased crime. Substantial portions of most Caribbean towns are run-down and neglected, with poor access to water and electricity reflecting the demise of the social environment. Directly linked to urban blight is the increase in uncontrolled urbanisation, particularly the recent growth of the informal sector as a result of social deprivation across the region.

The Informal Sector and the Caribbean Environment

For more than three decades, the problem of large-scale structural unemployment has been a central concern in the Caribbean as elsewhere in the developing world. According to Farrell (1978), Caribbean unemployment can largely be attributed to a shortage of capital, to the malfunctioning of the labour market, excessive rates of population growth, the importation of inappropriate technologies, and the effects of colonisation. Unemployment and underemployment figures for the region have been growing since the 1980s, with official unemployment exceeding 20 per cent and unofficial estimates reaching as high as 50 per cent (Deere *et al.*, 1990).

One outcome of the inability of many Third World countries to provide sufficient housing and employment for their growing labour force is the growth of the informal sector. The informal sector is playing a major role in providing employment within the debt ridden economies of many Third World countries (de Soto, 1989; Tokman, 1989; Portes and Castells, 1989), and this applies within the Caribbean as well (Thomas, 1988; Rampersad, 1991). In its present form, the informal sector refers to unaccountable and unregistered activities which are found in most countries of the world.

In the Caribbean context, the informal sector encompasses a range of retailing, production and service activities, including street trading (vending and higglering), craft production, smuggling, small-scale farming, food retailing, home-based clothing and electrical production, local household repairs and vehicle maintenance (Lloyd-Evans and Potter, 1992; Lloyd-Evans, 1994). Its merits, in contrast to formal employment, include independence, flexibility, and evasion of taxes and bureaucracy. In the Third World context, de Soto (1989) argues it represents a haven for motivated self employment and small business development, as well as jobs for the urban poor.

Despite its advantages in terms of employment creation, income generation and support for local communities in the Caribbean, it is apparent that there are a number of environmental issues associated with its prolific growth, particularly in urban areas. The informal sector is currently being supported by the Governments of Jamaica and Trinidad, through a series of income-generation schemes and training programmes. If this support continues, the environmental implications of rapid informal economic growth in terms of employment, agriculture, transport and housing, will have to be taken into account.

The nature of the informal sector, its evasion of laws and regulations, makes any restriction on growth difficult. In urban areas across the globe, the informal sector is increasing its hold on local communities. Unregistered and without adherence to planning controls, the growth of informal workshops, markets, transport systems and housing is resulting in a largely uncontrolled economy. Although the informal sector is providing necessary jobs, goods and services for low income communities, it is presenting a series of environmental hazards which are slowly increasing in severity.

Street Trading and Informal Markets: Urban Pollution and Congestion

Street hawking is a popular occupation in the region, especially in urban areas, and involves the retailing of a wide range of goods, from food to smuggled imported goods (Harrison, 1991). The traditional female 'higgler' or 'huckster' selling fruit and vegetables, or imported goods, is a familiar sight across the Caribbean (Katzin, 1960; Durant-Gonzalez, 1985; Le Franc, 1989). The importance of higglering in most countries of the Caribbean cannot be overestimated, as it plays an essential role in the food distribution system of many countries. For example, in Jamaica in 1975, a government minister concluded that 80 per cent of the fruits and vegetables consumed were sold by higglers (Senior, 1991). In Trinidad, East Indian traders have modernised the traditional role through the selling of fruit from stalls on the country's main highways. Street traders also sell a heterogeneous range of petty-commodities from jewellery and crafts, to imported cassettes and electrical items. Inter-island trading of commodities is also popular, and is an occupation often undertaken by women in the smaller islands of the eastern Caribbean (Phillips, 1985).

Despite the valuable contribution of vending or higglering to low-income communities in terms of employment and service provision, there are a number of negative factors associated with its operation. The trading of goods, from fresh vegetables to petty commodities, has long been regarded by governments as unaesthetic, but has often been tolerated due to its ability to provide incomes. According to Caribbean authorities, unauthorised vending is not strictly legal as it escapes detection, legislation and tax. In Trinidad, the police are given the power to arrest informal vendors engaged in 'wilful obstruction of the passage-way' or 'vending on the footpath', along with vendors displaying goods on railings, buildings, or littering the streets. In the recent past, the Trinidadian Government has launched many campaigns to clear the streets of Port of Spain and San Fernando of vendors.

The justification for periodic street clearance is centred around concern over litter, public hygeine, and street congestion. When the working day is finished, vendors will often discard perished fruit, paper and rubbish on the street, which then rots. Over time, large areas of refuse build up in residential areas. These areas often act as playgrounds, and a source of income, for many urban children who make money from collecting refuse or selling beer bottles back to the factories. Informal markets present a similar problem, but on a larger scale. Informally organised markets are common, and because they have no licence, they are not entitled to any kind of refuse disposal. Large piles of waste attract vermin, and eventually add to the degradation of the area. The problem of unauthorised waste disposal often extends to river dumping, and is leading to an increase in water-related disease in low-income residential areas.

A further problem, already noted by the Trinidadian authorities, is the poor level of hygiene found on the streets. For example, inter-island traders usually spend two or three days in one island, sleeping and cooking in market areas with no access to clean water or sanitation facilities. Similarly, street food sellers have no safe access to services or clean water, a fact which raises issues of public safety. Vending on the sidewalks of downtown Port of Spain or Bridgetown adds to the congestion of central urban streets, and haphazardly constructed stalls can prove to be dangerous to the public. The selling of 'mango-chow' or nuts at traffic-lights, has led to serious traffic accidents involving vendors, many of whom are children.

However, informal trading, which represents a livelihood to many of the urban poor, is on the increase in many islands due to escalating unemployment and economic recession. The environmental problems associated with informality are leading governments to take action to restrict its growth. Negative action against vendors will have social and economic repercussions for the urban population who make their living on the streets. Rather than pursuing street clearing operations, Caribbean governments need to take action to assist vendors to dispose of their waste, and to install services in markets and other public places. Assistance to render informal vending a more environmentally sensitive activity only requires a relatively small amount of planning and investment. Unfortunately, the environmental hazards associated with informal production present a more serious obstacle.

Petty-Commodity Production: Toxic Waste and Resource Decline

A potentially serious environmental problem is the expansion of informal production across the Caribbean, from small-scale industry and manufacturing units, to unregistered garages. The problem is particularly acute in the larger islands of Trinidad and Jamaica (Kirton and Witter, 1993; Lloyd-Evans, 1994). Haphazard development, usually without regard to safety and planning regulations, has led to the growth of dangerously constructed urban workshops and production units. In an attempt to emulate the formal manufacturing sector, many informal businesses have set up to produce low-cost goods either for local communities, or increasingly for export via foreign firms. The manufacturing of spare parts for vehicles, industrial chemicals and glues, clothing and electrical items are popular businesses in the region. Although a substantial proportion of petty-commodity producers are self-employed entrepreneurs, there is a trend across the Third World for larger companies to employ groups of informal workers to produce cheap goods for export.

In the 1990s, the use of subcontracted informal firms by large companies, often multinationals, means that increasingly, a vast range of goods is made in sweatshops, using old machinery and illegal practices (Safa, 1986, 1990; McAfee, 1991). Subcontracting is especially associated with the clothing and electronic industries, where low-paid female workers are located in small workshops using sub-standard materials, such as intoxicating glues or lead-based chemicals. These workshops often utilise cost-cutting practices and fail to conform to local legislation. Industrial effluents, containing non-biodegradable petro-chemicals, are frequently released from the unit into domestic water supplies, rivers or the sea. Waste products, including toxic metals and non-biodegradable substances, are disposed of on nearby waste grounds or discharged into local rivers, which may act as water supplies for nearby residents.

Non-conforming and dangerously constructed housing and workshops are all too frequent in the low-income urban communities of Kingstown, Castries, St Georges, Kingston and Port of Spain. As noted in chapter 5, the irony is that population and housing pressures are leading to increasing informal sector housing starts in hilly locations, just at the time when some are suggesting that climate change will result in fewer, but more intense storm events in the region (Potter, 1992d). This is particularly true of the hilly areas which make up the suburbs of Kingstown. On the eastern coast of St Vincent, as noted in chapter 6, hundreds of houses are located on the beach exposed to the full fetch of the Atlantic, the residents having been forced to move there by plantation owners (Potter, 1995). Informal garages, built in backyards, leave oil spillages to wash into the ground, and 're-tread' operators abandon tyres to rot in the sun. With no formal regulations, informal businesses discharge their waste into the atmosphere. Furthermore, the natural environment itself is often used as a resource by informal firms. Wood in particular is used by craftsmen or burned for fuel. Forest areas are becoming prey to informal suppliers, such as furniture makers, or more frequently, burned for fuel.

Despite these problems, however, the informal sector is potentially a more environmentally responsible economic system, as it represents an indigenous economy which utilises local resources and skills. Many informal activities, bottle, glass and paper collecting, for example, are already based on the principle of recycling. Across the Caribbean, old beer and soda bottles are collected and returned to the factory. In Trinidad, 'Vat 19' rum bottles are collected by small informal channa and confectionery producers, and used as containers for the sale of their products. In addition, the informal sector often uses more appropriate technology, local resources and operates on a smaller scale than large foreign businesses.

Attention does, however, need to be paid to assisting informal producers to dispose of waste and operate safer businesses, without adversely affecting their informal status. Public education could go a long way in assisting with the preservation of the local environment. The informal sector has brought benefit in terms of job creation and should be encouraged, but informal activities are likely to cause more environmental problems as the sector grows in size. This is becoming apparent in Trinidad, where uncontrolled urbanisation and the growth of the informal sector is resulting in visible degradation of the urban environment.

The Informal Sector and Environmental Degradation in Trinidad

The Republic of Trinidad and Tobago, a dual island state which maintains a varied pattern of social and economic activity, has witnessed a rapid growth in informal activity in recent years. Trinidad was fortunate to benefit from oil revenues in the 1960s and 1970s, but was hit by the oil crash in 1983 which threw the economy into rapid recession. The effects of international debt restructuring, a stagnant and dependent economy, escalating unemployment, rising urbanisation, urban primacy and increasing poverty currently places Trinidad in a vulnerable economic position. Environmentally, Trinidad displays a diverse ecosystem, ranging from the tropical forest of the northern coast to the mangrove swamps of Caroni and the South. The population is approximately 1.2 million with 60 per cent living in urban areas, and half a million living in Greater Port of Spain (see chapter 2).

Trinidad is the only Caribbean country with an important indigenous petroleum industry, and this has acted as the impetus for rapid industrial development and the establishment of oil refineries and energy-based industries (Scotland, 1983). Independence was gained in 1962, and the country became a Republic in 1974 under Prime Minister Eric Williams, who proceeded to direct investment towards heavy industry as the preferred path to development. A joint venture between foreign transnational companies and the government led to the ambitious development of the Point Lisas Industrial Complex in Southern Trinidad. It featured refineries, a chemical fertilizer plant, iron and steel works and smaller manufacturing industries. Between 1974 and 1983, the government was reported to have received US$ 17 billion in oil revenues, which was spent on industry, infrastructure and education. Large-scale urban construction, which included a major stadium and the famous twin tower complex, paralleled this industrial boom. Such prosperity was short-lived, however, as the unexpected fall in oil prices of the 1980s led to economic retrenchment. In 1988, Trinidad joined the ranks of those countries under the auspices of the International Monetary Fund (IMF), and found its economy controlled by an austerity package which included further cuts in public spending, public sector redundancies, the introduction of 15 per cent VAT and the recent opening up of the domestic market to imported luxury items. The official unemployment rate has remained in the region of 20 per cent since 1987, although the unofficial rate is thought to be nearer double that figure (Central Statistical Office, 1990). As a direct result, the country has experienced a growth in informal activity and unplanned settlement. The informal sector is one of the least documented sectors in the economy, but it is increasingly becoming one of the most important (Lloyd-Evans and Potter, 1992).

Consequently, the country is now being subjected to a new wave of environmental pressures which centre upon rapid urbanisation and unregulated business development, superimposed on an insufficient infrastructural base and limited space. In the past, Trinidad suffered from the government's preoccupation with the oil sector, sugar cane extraction, modernisation plans, and industrial development, all of which were instigated at the expense of the environment.

Current Environmental Pressures in Trinidad and Tobago: Degradation and Urban Land-Use

Throughout the 1970s, heightened public concern focused on the lack of environmental policy in the oil industry, particularly in the light of frequent oil spillages off the southern coast, and the leakage of poisonous effluents into the rivers and coastal zone. The south west and southern coasts of the country, particularly around Point Fortin, are subject to industrial development, the effects of which can be seen on the polluted beaches. The lack of strict environmental legislation in the formal economy, namely the petroleum and gas industry, is recognised by the government, but at present it is not the only environmental problem which the country is facing. The industrial modernisation trend of the 1970s and early 1980s led to rapid urbanisation and industrialisation in the north of the country, with the informal sector playing a major role in this development. This uncontrolled development is best exemplified along the burgeoning west-east urban corridor in the County of St. George, which extends from the capital, Port of Spain, to Arima in the east of the country (see Figure 2.10 for locations).

The 1982 National Physical Development Plan for Trinidad and Tobago provided a major comprehensive long-term overview of national development strategies to the year 2000. In relation to the environment, the plan states that "environmental planning and conservation essentially involve the dual and complementary activities of the allocation of land or physical resources to the most appropriate uses and the prevention or reduction of negative side-effects from positive development activities, which may result in the deterioration of the natural resource endowment and ecological imbalance" (National Physical Development Plan, 1982:74). Viewed in these terms, the Government admits that there is much room for the improvement of regulations, standards, administration and enforcement in the arena of environmental policy.

The 1982 plan also highlighted major environmental concerns in Trinidad, and it attempted to isolate the causal factors, as shown in Table 7.2. It is interesting to note that many of the problems are attributed to urbanisation and uncontrolled informal activity, and remain the subject of concern. In a recent development document, the Government stated that informal development was a major obstacle to urban environmental

**Table 7.2   Environmental problems associated with informal development
in Trinidad**

Nature of Environmental Problem	Causal Factors
Protection of fragile ecosystems and watersheds	Lack of legislation/ enforcement
Pollution on land and sea	Inadequacy/absence of control
Degradation of landscape	Inappropriate land-use (informal) Informal business development
Denudation of hillsides (erosion, flooding, blockage)	Haphazard building (informal) Small-scale agriculture Principal area: Northern Range, St George
Disposal of solid waste and effluents (high incidence of environmental related diseases, ie. gastroenteritis)	Industrial, commercial and domestic sources (mainly informal) in western periphery
Land-use conflicts Principal area: St George region	Intrusion of urban development

Source: National and Physical Development Plan of Trinidad and Tobago
(1982)

sustainability in the region (Medium Term Macro Planning Framework, Government of Trinidad and Tobago, 1991).

Informality and Unplanned Urban Development in Trinidad

The nature of informal activity suggests that even with improved Government control, the majority of informal activities will still avoid detection and legislation. Popular activities within the Trinidadian informal sector include farming, market gardening, building contracting, craft-production, shoemaking, clothing and electrical production, oil-related industry, housing and transport. In addition, small-scale distribution is popular, and encompasses petty traders, street hawkers, caterers of food and drink, bar attendants, agents and dealers. Other services include laundry and vehicle repair (Lloyd-Evans, 1994). Although valuable suppliers of employment, many of these activities are contributing to urban environmental degradation in Trinidad.

Uncontrolled development in Port of Spain is a major facet of informal retailing activity. Many residents feel that street trading in Independence Square, and informal developments such as the People's Mall on Frederick Street, are unaesthetic. A further example of uncontrolled development is trading in the burned out shells of the shops in downtown Port of Spain, which were damaged during the civil unrest of July 1990. As explained earlier, informal markets and the street trading of fruit and vegetables give rise to pollution from the lack of waste facilities. In Trinidad, trading is most prominent on Frederick Street in Port of Spain, the Eastern Main Road, Arima and San Fernando. The informality of the numerous street markets which operate along the Eastern Main Road is such that there are no sanitation facilities or organised rubbish clearance. Waste products are often left on the streets, or burnt in large fires.

The trading of fresh produce along the country's main highways is also popular, particularly amongst East Indian families. Although they provide a valuable service, highway traders constitute a hazard to motorists, and in the same way that they escape planning regulations, they also forfeit access to organised rubbish disposal. As a consequence, the highways are frequently littered with unwanted goods. Complaints from residents in the high-income suburb of Valsayn, relating to rotting food and rubbish along the Churchill-Roosevelt Highway, have led to violent street clearances of traders by the authorities in the recent past. Although statistics are not available, the incidence of highway accidents involving vehicles and traders is certainly on the increase.

Concern has also been directed at the growth of unplanned settlements on the hillslopes of the Northern Range. One of the primary causes of the denudation of the hillsides along the southern slopes of the Northern Range is the peripheral expansion of urban development in the capital, and in the adjacent county of St George. The rapidly consolidating urban corridor, which extends from Port of Spain through the commercial and residential suburbs of St George, is one of the fastest growing areas in terms of population, housing and informal employment, particularly in terms of home-based manufacturing, retail and subsistence agriculture. Increasingly, Trinidadians from the south and east are moving to the unplanned settlements along the urban 'east-west corridor' due to improved employment prospects and proximity to Port of Spain (Conway, 1989). Due to the pressure on restricted space, many families are building their own houses, small farms and workplaces on the surrounding hills of the Northern Range. This situation is giving rise to minor landslides and soil erosion, and it is a situation which is unlikely to change in the near future.

Urban development in the region has led to difficulties in water resource conservation, the destruction of forests, flash-flooding, and soil erosion, poor environmental conditions within the hillside settlements along the foothills, and the pollution of the San Juan River. Unplanned construction is also leading to the depletion of good agricultural land at the base of the forest. Furthermore, there is the increased use of the Northern forest for the poaching of rare species, as well as craft and furniture making by informal producers, an activity unrecorded by the government.

East Port of Spain, which is generally characterised by low-income settlements (see Potter and O'Flaherty, 1995), is particularly prone to rapid growth. Laventille and Sea Lots are the focus of industrial activity, which is undertaken in makeshift developments. This has resulted in the pollution of the Laventille Swamp, an area where local children play. The area directly below Lady Young Road, which consists of steep slopes and which is prone to landslides, had been zoned as a conservation area, but is now full of inroads in the mud which lead to more settlements.

Along the Eastern Main Road from San Juan to Arima, the landscape has become dotted with automobile junk yards, tyre re-tread centres, scrapyards and rubbish dumps. Informal small-scale production units are located in houses or back-yards in St Joseph and Tunapuna. These services range from welding, car repair, and shoe-making to small-scale furniture and appliance repairs. The regional sub-centres of San Fernando in the south, Arima in the east and Chaguanas in the central region, are also witnessing an explosion of informal activity. As previously discussed, petty commodity production and home-based industry avoid a number of emission laws and codes of safe practice. The haphazard development of repair shops and garages results in toxic waste being dumped into local rivers or the atmosphere. This is particularly dangerous when dealing with oil products, as heavy tropical rain often washes waste into the subsoil and residential water supplies.

The growth of informal developments is a major causal factor behind many of the environmental pressures recognised in Trinidad. This is particularly apparent in relation to the inappropriate use of the land, a problem which stems from the expanding development of irregular housing, and associated buildings on inappropriate land. The failure of the Government to cope adequately with this problem is reflected in the living environment, and in the incidence of environmentally-related diseases such as gastro-enteritis. The principal area where water borne infections are common is the western periphery, where population and economic activity are concentrated, and the inshore waters of the semi-closed Gulf of Paria, where wastes are discharged.

In this respect, alternative solutions need to be adopted. There is a pressing need for environmental education and awareness to highlight the effects of human activity in small ecosystems, and the value of preserving untouched areas. Conservation areas have already been designated in the Northern Range, part of the Central Hills, and in marine areas where coral reefs exist, but there are a number of bureaucratic problems associated with the implementation of such policies. The structure of administration is such that development policies and responsibilities fall between a number of separate ministries, which frequently operate independently of one another. In most cases, the respective agencies are solely responsible for a resource, such as WASA, which administers water and sewerage. This leads to a fragmentation of development policies and inconsistencies regarding goals.

The Future of Environmental Policy in Trinidad: The Informal versus the Formal Sector and the Green Debate

In the context of recent national development (Medium Term Macro Planning Framework, Government of Trinidad and Tobago, 1991), the Government of Trinidad and Tobago has highlighted a number of environmental objectives. Firstly, there is the development of a high quality living environment, which includes the raising of health standards, in a context of effective resource development which retains the aesthetic value of the landscape. This goal has direct implications for the eradication of some informal activity, particularly street trading. Secondly, there needs to be an increased contribution to economic welfare, through the informed use of resources, whereby the high cost of remedial infrastructural works and environmental programmes can be channelled into productive expenditure, and the opportunities for natural growth are not reduced by destructive activity.

In order to achieve this, the Government intends to focus on a number of specific targets. Firstly, the Government wants land to be allocated to its optimum use, which in many cases will be economic. Secondly, it wants to preserve valuable ecosystems, such as the Northern forest and the mangrove swamps. Thirdly, it aims to cope with the negative effects of development such as pollution, and encourage environmental awareness and appreciation among the population. Fourthly, it aims to establish effective control of informal urban growth by adherence to national planning, with the introduction of new controls where necessary. Particular attention focuses upon the illegal development of land, and the denudation of vegetated and marine areas. An additional idea centres on establishing Green Belts around the capital to ease developmental pressure.

Despite a number of problems associated with the development of the informal sector, there are positive features of this activity which favour it in comparison with more formal development. Firstly, the informal sector is small-scale, local and more indigenous, when compared with the large-scale, formal economy, which is subject to external control. The local worker has greater links with her or his environment, and in this respect, should consider taking better care of resources. Many of the environmental problems associated with the informal sector are relatively easy and cheap to address, such as improved rubbish disposal. There are also many activities which involve recycling, and there is less waste of resources from advertising and packaging. Attention needs to be paid to using appropriate, although environmentally safe, technology and natural resources, such as vegetable dyes in the clothing industry.

Generally, there is need for improved water and sewerage infrastructure, and the local control of rubbish pollution. Such improvements will be difficult to implement due to the unregistered nature of the activities, as this type of targeting is a method of registering informal activities, thereby making them liable to tax, laws and other regulations. There could, however, be the provision of more municipal dumping grounds for rubbish in the city, and a general improvement of basic infrastructure.

Unfortunately, by advocating the growth of the informal sector, Third World governments are giving way to uncontrolled development and urban degradation. This is the dilemma which is currently facing the Government of Trinidad and Tobago as it tries to balance the negative and positive effects of the sector for the future development of the country. Possible solutions lie in educating the small-business entrepreneurs and operators about the environment. Local entrepreneurs should be more concerned with the future of their own country, than outside interests. In attempting to encourage the informal sector, the Trinidadian government is making training courses and loans available to individuals. Environmental awareness should form a part of training courses, and loans should be subject to following certain basic environmental guidelines. Essentially, the informal sector can in theory present an eco-friendly form of local development.

Conclusions: Balancing Caribbean Development with Environmental Protection

This chapter has attempted to highlight many of the environmental problems which are associated with current patterns of urban and industrial development in the Caribbean. However, as shown by the Rio Summit, environmental concerns are often pushed aside by Third World countries due to pressing short-term concerns, such as poverty alleviation and debt repayment. In such instances, the balance between environmental safety and economic progress is usually unequal, with the environment very evidently taking second place. In other words, all too often, the dictates of the global economic system override local circumstances and concerns.

One such example is the current growth of the informal sector in the region. The ability of the informal sector to avoid legislation can be beneficial in terms of employment creation, but its uncontrolled nature can cause many problems to planners and environmentalists. As physical planning policies often fail, education could play an important role in improving environmental quality in the Caribbean. Improved environmental awareness could be a first step to solving current problems. What is needed is a method of transmitting alternatives to environmental degradation to the population, whilst other avenues of economic activity are found. For example, it is difficult to instruct vendors of sea-coral to stop trading if they have no alternative source of income.

In addressing these problems, the policy makers need to readdress legislation. The last thirty years have probably seen higher levels of environmental degradation in the Caribbean than in the rest of its history. There needs to be another path to economic development which halts the exploitation of natural resources. In the present political environment, environmentally sustainable local economies could provide a better form of development for the Caribbean. Unless the adverse affects of informal development are given due consideration, however, urban environments may well continue to degrade.

8 Caribbean Urban Futures

ROBERT B. POTTER and JONATHAN PUGH

The chapters of this monograph have already served to demonstrate that the future of the Caribbean region is intimately tied up with both the urban and the global. Indeed, a central theme throughout has been that this conflation of urbanity and the global has been central to the region since its first urban settlements were established. The relationship became stronger and stronger with the subsequent colonial control and domination of Caribbean territories. The second major theme, therefore, is a direct corollary of these circumstances, for both socially and environmentally, the local has tended to be persistently overlooked and undervalued in the Caribbean region, due to the hegemony of the global and the international.

However, the aim of this final concluding chapter is not to revisit these themes. Rather, the remit is to consider the implications of such circumstances for sensitive and appropriate planning and development within the region. In this context, planning and development may be defined as overseeing and guiding future change toward some prescribed set of societal objectives or goals (Hall, 1974; Potter, 1985). As such, planning is all about establishing desired futures – both rural and urban. However, following this line of argument, it is all too easy to see planning as the solution to all of society's problems. This view of planning as the amelioration of everyone's difficulties is obviously naïve and simplistic, not least because somebody's loss is likely to be someone else's gain. Seeing planning as a cure-all for societies' ills tends to imply that planning is a straightforward and technocratic process – a set of clear procedures, which if followed and implemented correctly, will lead inescapably to a better world for all. However, given sectional and class interests, whose goals will be implemented first and foremost? In short, who will gain and who will loose, given that in the short-term, at least, societal change tends to be a zero sum game?

In chapter 3, the argument was presented that in the post-war period, planning and development in the Caribbean have tended to be top-down in both conception and implementation. In seeking to develop, most Caribbean nation states have followed the path to industrialisation that had been adopted in the west. As detailed in chapter 3, several Caribbean nations followed directly, the prescriptions of Sir Arthur Lewis. In addition, since the 1960s, tourism has been taken as the second major plank of development. As shown in the early chapters, this has served to tie the region yet more closely into the global capitalist system. Many commentators have talked about the period of colonial dependency merely being replaced by one of acute neo-dependency.

It is all too easy to say that planning should embrace a bottom up philosophy, and that planners must open up two-way channels of communication, between themselves as the so-called experts, and those for whom they are doing the planning. Thus there is a paradigm that is growing in popularity in both academic and policy-oriented circles that argues that planning must be people-based and participatory. But how exactly is this to be achieved?

In the Caribbean, as elsewhere in the Third World, one of the issues to be addressed is that many planners have received their academic and professional training abroad. Frequently, they have studied at one or two leading universities in the United States or the United Kingdom, with these often being located in the principal metropolitan centres. Zetter (1981) argues that this can give rise to what he refers to as "cultural collisions" between the environmental needs of developing countries on the one hand and the conventional wisdoms and accepted solutions of decision-makers on the other. These may occur principally via the adoption of imported, inappropriate and frequently over-sophisticated models and modes of planning. This is likely to occur in two main fashions. Firstly, planners may come to regard Westminster-style solutions and standards as the norm and apply them in extremely divergent cultural, social and environmental circumstances, with insufficient critical appraisal of their true worth. Second, there may be a predisposition to adapt complex techniques for the apparent sophistication and scientific objectivity which they appear to bestow upon the user (Potter, 1985). Further, Lloyd (1979) notes that having been trained abroad, planners become even more of an elite, and may unwittingly take an increasingly pejorative view of those they are employed to serve, or at the very least, become distanced from them.

An example of the adoption of an over-sophisticated set of urban planning norms is exemplified by the first Physical Development Plan for Barbados, produced in 1970 (Government of Barbados, 1970). This presented what can only be described as a highly ambitious programme for the decentralization of employment, population and government offices into an almost perfect Christaller-type central place hierarchy of the sort described at the beginning of chapter 2. The plan has been discussed at length in Potter and Hunte, 1979 and Potter, 1985. As shown in Figure 8.1, the strategy envisaged the growth of a whole series of central places spanning from the village, to the district and regional-levels. Thus, small rural areas such as St John were forecast to become district-level service centres, in this instance, housing a population of 5,000.

Although this first national settlement strategy was almost totally idealistic, and was certainly never implementable as presented, the second and third national Physical Development Plans, produced in 1983 and 1998 respectively, envisaged spatially more-focussed decentralisation along the linear urban corridor (Government of Barbados, 1983; 1998). In this way, the local east-west impact of Speightstown, Holetown, Oistins and Six Cross Roads as service centres was to be enhanced (see Potter, 1986 and Pugh and Potter, 1999). In these two plans, therefore, as can be perceived in Figures 8.2 and 8.3, decentralisation and urban regeneration were being planned in a fashion that was cognizant of the pre-existing constraints implied by the extant coastal-mercantile urban system.

In the case of this example, it is not the intention to develop a central-place-type settlement system that is at fault, but rather the manner in which this was envisaged in relation to the pre-existing settlement structure. In direct contrast, the effort to create new urban places in post-revolutionary Cuba affords an example of matching carefully a set of planning proposals with the existing socio-economic realities and the political prescriptions adopted. At the inter-regional scale, the growth of provincial towns in the 20,000 to 200,000 size range was stimulated in Cuba in an effort to counterbalance the primacy of Havana (see Figure 8.4). At the next level down, the regrouping of villages into rural towns was effected (Hall, 1981). In the fields of both health and education, massive efforts have been made to develop new secondary schools along with primary, secondary and tertiary health care facilities, in what were predominantly rural areas. Thereby, as noted previously in this monograph, the policy aim has been to ruralise the towns and to urbanise the countryside.

On the other hand, it can be argued that however difficult to achieve, participatory planning is of particular salience with regard to developing countries. Colonialism can be interpreted as a phase when democratic rights were systematically repressed. Thus, wider representation is strongly needed in independent nations of the Third World. There have always been strong reasons for encouraging citizen participation, in that people should be more strongly behind plans if they are party to their initial formulation (Conyers, 1982). At the same time, with states being rolled

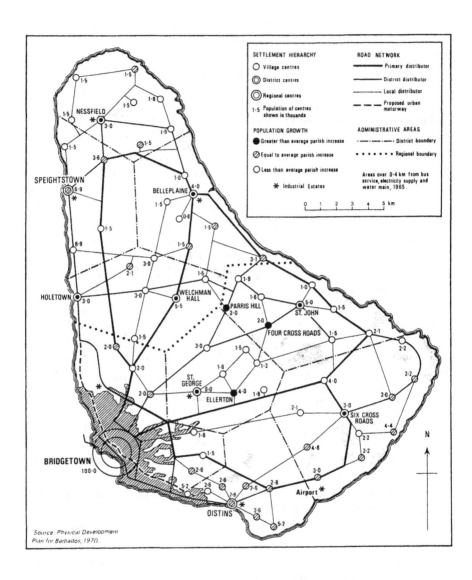

Figure 8.1   The settlement configuration envisaged in the first Physical Development Plan for Barbados

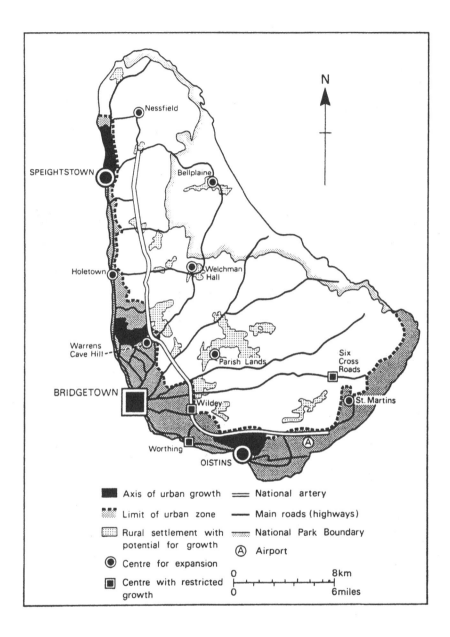

Figure 8.2   The principal proposals of the 1983 Revised Physical Development Plan for Barbados

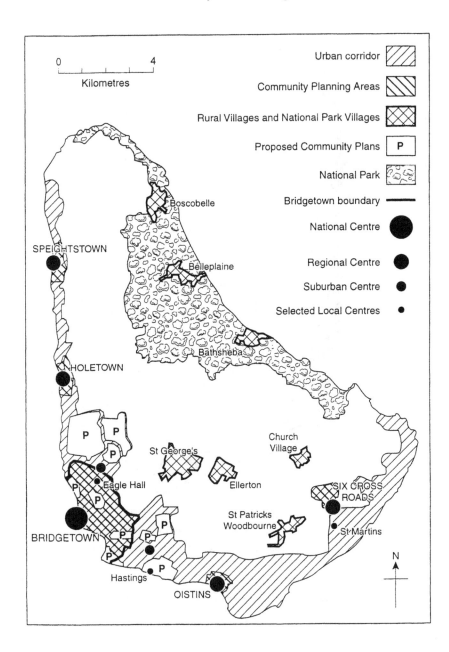

Figure 8.3　The principal proposals contained in the 1998 Draft Physical Development Plan for Barbados

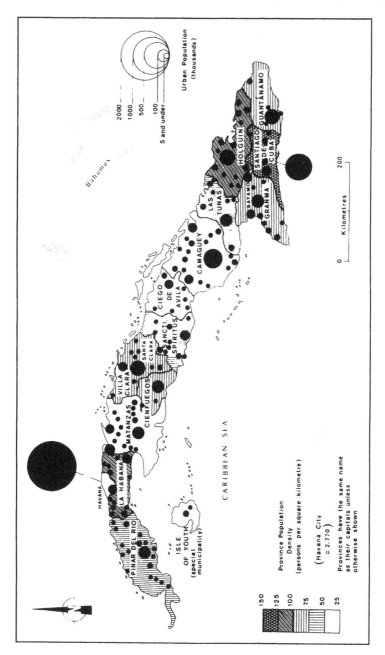

**Figure 8.4   The distribution of Cuban urban settlements by the mid-
1980s**
(Source: after Hall in Potter, 1989)

back under structural adjustment programmes, it is axiomatic that members of the general citizenry are bound to be expected to be more and more involved in the development process on a day-to-day basis. To some extent this simple fact explains why international organisations such as the World Bank are now espousing the need for public participation and empowerment.

This is not to say, of course, that Caribbean states have never consulted their citizenry on planning issues. An example is provided by the Physical Development Plan produced for Trinidad and Tobago in the early 1980s (Government of Trinidad and Tobago, 1982). The significance of the plan is that it suggested, and offered for evaluation, four alternative strategies that could be employed to respond to the strong urban concentration that has developed from the colonial period. The four strategies are shown in Figure 8.5 and ranged from trends development of the existing urban areas of Port-of-Spain and San Fernando-Couva, to the complete dispersion of new development. In the end, the fourth compromise solution was adopted, which envisaged the combination of dispersion with concentration. Growth was to be concentrated into a number of localities, such as Sangre Grande and Couva-San Fernando, but was also to be spread to more remote areas such as the rural south and south-east (see Figure 8.6). However, although the plan identified these alternatives, the procedures put in place for eliciting the publics' views were limited. But this need not necessarily be the case.

In the context of the urban Caribbean, recent developments seem to suggest that within the region there are genuine possibilities for the adoption of 'locally-defined' paths to development. Such developments suggest that a more democratic, decentralised, iterative and integrated approach toward development planning could be encouraged, in order to achieve more participatory – and therefore effective and sustainable - development. This approach is the reverse of the top-down, non-participatory development approach that the region has hitherto generally experienced.

In particular, chapter 3 highlighted how global trends have meant that Caribbean societies have 'converged' on other countries in terms of their expectations. However, this chapter also indicated that societies have 'diverged' because of the limited ability of many communities to realise such goals, and the highly skewed production possibilities that they face. One possible means of counteracting such trends is to attempt to foster more 'locally-defined' development imperatives. In essence, this means including more people in the development process on a continuous basis.

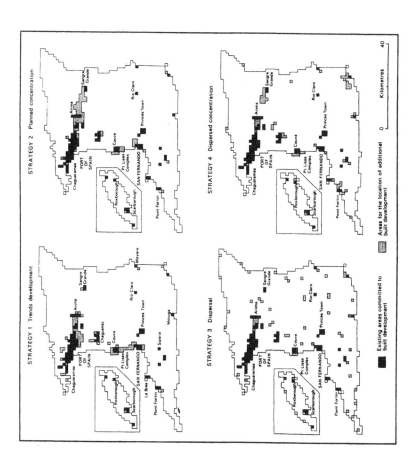

Figure 8.5   Alternative plans for urban and rural settlements in Trinidad and Tobago

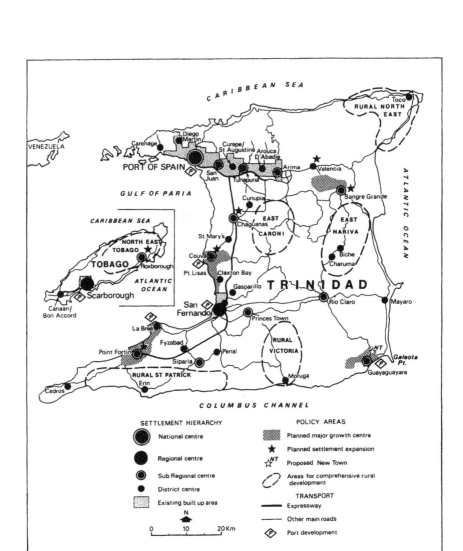

Figure 8.6   The urban and regional development plan adopted for Trinidad and Tobago

This it may be argued should result in the 'resources cake' being more equally divided, and a greater concern being expressed for the urban environment.

In fact there has been a veritable plethora of conventions and conferences supporting the ideal of equitable sustainable development through public participation in decision-making processes. In relation to the Caribbean region, a number of key documents have been produced for the region as a whole, as well as by individual countries. This was principally brought about in preparation for the Rio Earth Summit of 1992, and in order to assist with the subsequent implementation of Agenda 21. One of the central themes in all of these documents is the need to increase the institutional capacity of local organisations so that they can include people more directly in the decision-making process. For example, *Our Own Agenda*, produced by the Latin American and Caribbean Commission on Development and the Environment (1990:75), states that: "The point is, first of all, to undertake changes of a political and institutional nature, designed to modify the state-society relationship so that the state may actually become the interpreter of society's objectives".

It is widely acknowledged that the first group of countries to significantly respond to Agenda 21 were the small island developing states (SIDs). A conference was held in Barbados in 1994 in order to detail what Agenda 21 meant to these countries, given the nature of the social, economic and environmental problems they face. The resulting *Small Island Developing States Programme of Action* (SIDs POA) highlights the need to "increase the awareness and involvement of non-governmental organisations, local communities and other major groups in public education, national planning and the implementation of sustainable development programmes" (United Nations, 1994:29). Capacity 21, a programme which operates in six Caribbean islands, and which was produced in order to implement Agenda 21 and the SIDs POA more effectively, "seeks to strengthen the capacity of national and regional institutions to implement sustainable development plans and programmes" (Caribbean Centre for Development Administration, 1997:2). Those countries involved in implementing Capacity 21 – Barbados, the British Virgin Islands, Dominica, Grenada, Jamaica and St Lucia – have all established some type of Sustainable Development Council. This is a national coordinating mechanism that is designed to involve different sectors in the production and implementation of a sustainable development strategy. Thus, many Caribbean countries have signed up to the idea of collaborative environmental management. The implications for the apparent change in approach towards development for the urban Caribbean will now be addressed.

There are several "beacon" schemes which represent a tentative movement towards more 'people-centered' development in the urban Caribbean, albeit on a somewhat limited and constrained basis. Innovative, 'locally-defined' institutional mechanisms for environmental management have been introduced in several urban areas since the Rio Summit was held in 1992. The success of this approach is increasingly being acknowledged, with the consequence that the central theme of 'inclusive development for sustainability' is slowly being adapted, at least in some parts of the Caribbean. However, given the long history of top-down politics, government and development in the region, undoubtedly this should be seen as a tendency, albeit one with much future potential.

Two grassroots planning experiments illustrate this line of argument, as well as exemplifying the serious constraints which are to be faced. These are firstly, the *Soufriere Marine Management Area* (SMMA) on the west coast of St Lucia; and secondly, the *Folkestone Marine Park and Reserve* (FMPR) on the west coast of Barbados. The former has a longer history of inclusive environmental management than the latter. Indeed, those who have recently been responsible for producing new procedures for managing the Folkestone area have adapted many of the approaches originally developed in St Lucia. The salient issues in both areas are the conflict within the littoral area between tourism and local activities such as fishing, plus the lack of involvement of stakeholders in development in the past. In both cases, the situation has almost come to physical conflict and the response has been to establish roundtables and committees involving a range of stakeholders, in an effort to suggest the ways in which these areas should be managed. This situation has been exacerbated due to the intensive development on the coastlines of Caribbean countries, as detailed in chapter 2.

The SMMA formally began in 1995, with the production of the Soufriere Marine Management Area Plan. Soufriere is the third largest urban area in St Lucia, with approximately 8,000 inhabitants. Before the final agreement for the management of resources was produced, there had been at least three years of intensive consultations, with the process being supported by organisations such as the Environmental and Coastal Resources Project and the French Mission for Technical and Cultural Cooperation. The process is now largely self-funding, with most of the operating costs coming from the diving and yachting sectors. In general, the environmental impacts of the SMMA have been to reduce degradation on the reefs, and in some cases, to arrest entirely such deleterious change.

The FMPR was established through legislation in the early 1980s. However, the lack of enforcement of the legislation and the minimal involvement of local stakeholders influenced the design of a $2 million project aimed at increasing local peoples' involvement in the development of three intensively used areas – Carlisle Bay, Harrisons Cave and the FMPR. This project has only been running since 1998 and the procedures to be employed in respect of environmental management are still being finalised. In Soufriere, a Technical Advisory Committee meets quarterly and is made up of stakeholders from different sectors. This decides upon the management of most of the Soufriere coastal area. The process is soon going to be extended farther to the north, to cover Canaries and Anse la Raye. Many of the stakeholders in Folkestone, observing the success of the SMMA, have called for a similar committee to be established in Folkestone.

The support given to these two processes by the Governments of Barbados and St Lucia suggests a willingness to establish 'locally-sensitive' institutions. For example, the establishment of the SMMA was predicated on the *Fisheries Act 1984*, and the Barbadian Government provided some funds to hire a Canadian consultant in order to produce a report to assist the management of the FMPR. However, there are issues that influence these attempts at 'locally-defined' development. Firstly, the focus is very much tourism-based, which in turn, is a reflection of the dominance of global capitalism. Does this matter? Can 'locally-sensitive' development still occur even under such circumstances? These questions can be answered by discussing the specifics of the project processes in more detail, along with the wider frameworks for development in these two countries.

Firstly, in both cases, the real impact of the local fisher people on change is debatable. In Soufriere, there are only usually two or three representatives from the fishing industry on the Technical Advisory Committee, with the majority of members coming from the tourist sector. Secondly, the Community Development Plan for Soufriere has a very strong tourist focus and thus the direction from Government is clearly that Soufriere is to be developed through tourism. A similar situation can be recognised in Barbados. Although on a day-to-day basis hoteliers are not vocal at the meetings of the roundtable, this has been interpreted by some as a reflection of the greater actual influence that they have at the national level, and as indicating that 'locally-defined' development is less relevant. Also at this national level, the new draft Physical Development Plan for

Barbados (1998) clearly envisages the continuation of this area as a 'tourist zone'. Indeed, this is the perception of many of the members of the Folkstone roundtable and has resulted in resentment of the process amongst some. Many fisher people in particular believe that their inclusion is 'tokenistic' and is merely an attempt by Government and the tourist sector to keep them on their side. Furthermore, the lack of involvement of town planners on the roundtable suggests that 'locally-defined' development is not seen as an important issue in physical development planning for the area. Indeed, a lack of public participation appeared to characterise the preparation of the third Physical Development Plan discussed previously in this chapter. Thus, these developments show both the opportunities and the constraints which are to be faced in implementing bottom-up approaches to planning – in particular, the hegemony of tourism-oriented urban genesis, as argued in chapter 2.

Indeed, many national institutions and organisations are not adopting the inclusive approach needed to support the SMMA and FMPR. This is perhaps best illustrated in the case of the Sustainable Development Councils. As indicated earlier, these have been established in Barbados and St Lucia to implement Capacity 21, and to provide support for more participatory approaches toward managing the urban environment. However, these councils have been viewed as 'one off projects' rather than as components of continuing processes. In both territories, it is widely acknowledged that the Sustainable Development Councils are unlikely to meet again. Furthermore, the Councils have tended to exclude certain environmental non-government organisations, as well as the public, from their activities. Thus, the process has begun to be seen as one of 'non-decision-making', with the real decisions remaining at the governmental level. Thus, the 'collaborative' and 'continuous' elements needed to support 'people-centred' development are largely missing.

Whilst many of those in government acknowledge the need for an overall coordinating mechanisms which could support more 'locally-defined' urban development, the realities of global capitalism and the impact which this has on national policy-making further restrict such an approach. Innovative mechanisms at both the national and local levels therefore remain isolated. It is difficult to build upon such processes when the 'lure of the fast buck' is so strong and pervasive.

The above discussions suggest that both the SMMA and the FMPR are not serious attempts at the decentralisation of power and may more readily be interpreted as reactions to the impact of global capitalism, and especially, coastal-based tourism. That is to say, they are primarily concerned with mediating the impacts of a national policy – which is a manifestation of global capitalism - than with creating institutions to support 'locally-defined' development. Thus, areas such as Folkstone and Soufriere become less likely to develop integrated approaches to development, but continue to develop 'local specialities' within the process of global divergence. These two examples therefore support the notion that the primacy of Bridgetown in Barbados, and Castries in St Lucia are highly likely to be maintained. To conclude, these examples and the others contained in this book, suggest that the primacy of the urban littoral area is likely to continue in most Caribbean territories.

Indeed, speaking in these terms, we return once again to the central theme of this volume, that the towns and cities of the Caribbean must be placed in the wider context of the evolution of the global urban system. In this connection, it has recently been argued that Miami should be regarded as the capital city of the Caribbean region (Grosfoguel, 1995). Since the 1960s, Miami has grown rapidly as a haven for retirement and as a centre for tourism. However, since the 1970s, the city has grown as an international banking and trade centre for the entire Caribbean basin, reflecting the relative decentralisation of these activities from New York and other large northern cities. Thereby, Miami may be viewed as a secondary World City, which exercises capital management and control functions for the entire Caribbean region. This nesting within the global urban hierarchy attests to the likelihood of enhanced urban influence in the future, and the enhancement of processes such as global convergence and the demonstration effect, as discussed earlier in the present text.

A final example of this latter point relates to existing urban densities in the Caribbean and the likelihood of high-rise urban developments. Many of the smaller urban places of the Caribbean are noticeably low-rise, with the majority of buildings not extending much above two or three stories. This is certainly true of Bridgetown, Barbados, Castries, St Lucia and Kingstown, St Vincent. Given the small size and limited land availability which is characteristic of these territories, this may be an interesting area where urban change is likely to occur in the future. Indeed, there are some indications of incipient high-rise development in these countries. Firstly, this is occurring with respect to symbolic high-rise government office blocks, such as the twin towers of Port of Spain and the government and commercial blocks of Castries. However, it is tourism together with Americanisation and the demonstration effect that seem to be the key influences in relation to the occurrence of high-rise urban developments, as exemplified by the Condado area of San Juan, Puerto Rico.

We finish, therefore, by returning to our central recurrent theme, namely that the nature of urban development in the Caribbean region is, and always has been, closely linked with the global and the international.

Bibliography

Acosta, M. and Hardoy, J. (1973), *Urban Reform in Revolutionary Cuba*, Occasional Paper No.1, Antilles Research Program, Yale University, New Haven, Connecticut.

Albuquerque, K. and McElroy, J. (1989), 'Puerto Rico and the United States Virgin Islands', in R.B. Potter (ed), *Urbanization, Planning and Development in the Caribbean*, Mansell, London and New York.

Allcock, J. (1992), 'International tourism and the appropriation of history', paper presented to the conference Tourisme International entre Tradition et Modernité, Carrefour Universitaire Méditerranéan, Nice, 19-23 November.

Amato, P. (1970a), 'A comparison of population densities, land values and socio-economic class in four Latin American cities', *Land Economics*, vol.46, pp.447-455.

Amato, P. (1970b), 'Elitism and settlement patterns in the Latin American city', *Journal of the American Institute of Planners*, vol.36, pp.96-105.

Armstrong, W. and McGee, T.G. (1985), *Theatres of Accumulation: Studies in Asian and Latin American Urbanization*, Methuen, London and New York.

Augelli, J.P. and West, R.C. (1976), *Middle America: Its land and Peoples*, Prentice Hall, Englewood Cliffs.

Bacon, P.E. (1995), 'Wetland resource rehabilitation for sustainable development in the eastern Caribbean', in D. Barker and D.F.M. McGregor (eds), *Environment and Development in the Caribbean: Geographical Perspectives*, University of the West Indies Press, Kinston, pp.46-56.

Barbados Advocate (1993a), 'Crop Over Programme all set', 15 April.

Barbados Advocate (1993b), 'Little Trinidadian scores "big bucks" with Coca Cola', 27 April.

Barbados Advocate (1993c), 'Tourism unity a must', 26 July.

Barbados Advocate (1993d), 'Castle takes on a new look', 27 September.

Barbados Advocate (1993e), 'Lewis encourages cultural growth', 3 November.

Barbados Advocate (1993f), 'Duty free push start', 25 November.

Barbados Advocate (1993g), 'Advertisement for heritage fair', 2 December.

Barbados Advocate (1993h), 'Christmas cheer at Simpson Motors', 3 December.

Barbados Holiday Map (1991), *Barbados Holiday Map*, December 1990 to May 1991, Paperchase Publications, Bridgetown.

Barbados Wildlife Reserve (no date), *Barbados Wildlife Reserve* publicity material.

Barrow, C. (1992), *Family Land and Development in St Lucia*, Institute of Social & Economic Research, Bridgetown, Barbados.

Barry, T. (1992), 'The great mall of China', *Toronto Star*, 1 February.

Barry, T., Wood, B. and Preusch, D. (1984), *The Other Side of Paradise: Foreign Control in the Caribbean*, Grove Press, New York.Beavon, K. (1977), *Central Place Theory: A Reinterpretation*, Longman, Harlow.

Beckford, G.L. (1972), *Persistent Poverty: Underdevelopment in Plantation Economies of the Third World*, Oxford University Press, New York.

Beckford, G.L. (1975), 'Caribbean rural economy', in G.L. Beckford (ed), *Caribbean Economy: Dependence and Backwardness*, Institute of Social and Economic Research, Mona, Jamaica.

Berry, B.J.L. (1972), 'Hierarchical diffusion: the basis of development filtering and spread in a system of growth centres', in N.M. Hanson (ed), *Growth Centres in Regional Economic Development*, The Free Press, New York.

Besson, J. (1987), 'A paradox in Caribbean attitudes to land', in J. Besson and J. Momsen (eds), *Land and Development in the Caribbean*, Macmillan, London, pp.13-45.

Best, F. (1993), 'Turst fair truly nostalgic', *Sunday Advocate*, 12 December.

Blume, H. (1974), *The Caribbean Islands*, Longman, London.

Boorstin, D. (1987), *The Image: A Guide to Pseudo Events in America. 25th Anniversary Edition*, Atheneum, New York.

Boswell, T. and Biggs, J.E. (1989), 'The Bahamas', in R.B. Potter (ed), *Urbanization, Planning and Development in the Caribbean*, Mansell, London and New York.

Brierley, J. (1985), 'A review of development strategies and programes of the People's Revolutionary Government in Grenada, 1979-83', *Geographical Journal*, vol.151, pp.40-52.

Britton, R. (1980), 'Let us handle everything. The travel industry and the manipulation of the travel experience', *USA Today*, May, pp.45-47.

Britton, R. (1994), 'Tourism and transport', *Guyana Review*, March, pp.20-23.

Brookfield, H. (1975), *Interdependent Development*, Methuen, London.

Broom, L. (1953), 'Urban research in the British Caribbean: a prospectus', *Social and Economic Research*, vol.1, pp.113-119.

Browning, H.L. and Roberts, B.R. (1980), 'Urbanization, sectoral transformation, and the utilization of labor in Latin America', *Comparative Urban Research*, vol.3, pp.86-103.

Bryden, J. (1973), *Tourism and Development: A Case Study of the Commonwealth Caribbean*, Cambridge University Press, New York.

Buck, R. (1978), 'Boundary maintenance revisited: tourist experience in an old order Amish community', *Rural Sociology*, vol.43(2), pp.221-234.

Burgess, R. (1982), 'Self-help housing advocacy: a curious form of radicalism: a critique of the work of J.F.C. Turner', in P.M. Ward (ed), *Self-Help Housing: A Critique*, Mansell, London.

Burgess, R. (1990), 'The state and self-help building in Pereira, Colombia', unpublished PhD thesis, University of London.

Burgess, R. (1992), 'Helping some to help themselves: Third World housing policies and development strategies', in K. Mathéy (ed), *Beyond Self-Help Housing*, Mansell, London, pp.75-91.

Butler, R. (1980), 'The concept of a tourism area cycle of evolution. Implications for the management of resources', *Canadian Geographer*, vol.24, pp.5-12.

Captain Morgan Cruises (1993), *Captain Morgan Cruises, Winter 1994/95 Programme*, Commercial Advertising Services, Valetta.

Caribbean Development Bank (1992), *Annual Report 1992*, Bridgetown, Barbados.

Cazes, G. (1976), 'Le tiers-monde vu par les publicités touristiques: une image géographique mystifiante', *Cahiers du Tourisme*, série C, no.33.

Cazes, G. (1987), 'L'île tropicale, figure emblématique du tourisme international', *Cahiers du Tourisme*, série C, no.112.

Central Statistical Office (1990), *Annual Statistical Digest, Republic of Trinidad and Tobago*, Port of Spain.

Christaller, W. (1966), Die Zentralen Orte in Suddeutschland, Verlag, translated by C.W. Baskin (1966), *Central Places in Southern Germany*, Prentice-Hall, Englewood Cliffs.

Clarke, C. (1974), 'Urbanization in the Caribbean', *Geography*, vol.59, pp.223-232.

Clarke, C. (1986), 'Sovereignty, dependency and social change in the Caribbean', in *South America and the Caribbean*, Europa Publications, London.

Clarke, C. (1989), 'Jamaica', in R.B. Potter (ed), *Urbanization, Planning and Development in the Caribbean*, Mansell, London and New York.

Clayton, A. and Potter, R.B. (1996), 'Industrial development and foreign direct investment in Barbados', *Geography*, vol.81, pp.176-180.

Cohen, E. (1979), 'A phenomenology of tourist experiences', *Sociology*, vol.13, pp.179-201.

Cohen, E. (1982), 'The Pacific islands: from utopian myth to consumer product', *Cahiers du Tourisme*, série C, no.27.

Cohen, E. (1985), 'Tourism as play', *Religion*, vol.15, pp.291-304.

Cohen, E. (1986), 'Tourism as time', *World Leisure and Recreation*, vol.28(3), pp.13-16.

Cohen, E. (1988), 'Authenticity and commoditization in tourism', *Annals of Tourism Research*, vol.15, pp.371-386.

Cohen, E. (1993), 'Contemporary tourism – trends and challenges', paper presented to the International Academy for the Study of Tourism, Seoul, 23 June.

Conway, D. (1981), 'Fact or opinion on uncontrolled peripheral settlement in Trinidad: or how different conclusions arise from the same data', *Ekistics*, vol.286, pp.37-43.

Conway, D. (1982), 'Self-help housing, the commodity nature of housing and amelioration of the housing deficit: continuing the Turner-Burgess debate', *Antipode*, vol.14, pp.40-46.

Conway, D. (1983), 'Tourism and Caribbean development', *Universities Field Staff International Reports*, no.27, 12pp.

Conway, D. (1989), 'Trinidad and Tobago', in R.B. Potter (ed), *Urbanisation, Planning and Development in the Caribbean*, Mansell, London, pp.49-76.

Conyers, D. (1982), *An Introduction to Social Planning in the Third World*, Wiley, Chichester.

Cooke, P. (1983), *Theories of Planning and Spatial Development*, Hutchinson, London.

Crick, M. (1985), 'Tracing the anthropological self: quizzical reflections on fieldwork, tourism and the ludic', *Social Analysis*, vol.17, pp.71-92.

Crick, M. (1989a), 'Representations of international tourism in the social sciences: sun, sex, sights, savings and servility', *American Review of Anthropology*, vol.18, pp.307-344.

Crick, M. (1989b), 'The hippy in Sri Lanka. A symbolic analysis of the imagery of schoolchildren in Kandy', *Criticism, Hersey and Interpretation*, vol.3, pp.37-54.

Crick, M. (no date), 'Tourists, locals and anthropologists: quizzical reflections on "otherness" in tourist encounters and tourist research', draft manuscript.

Cross, M. (1979), *Urbanization and Urban Growth in the Caribbean: An Essay on Social Change in Dependent Societies*, Cambridge University Press, London.

Cynthia (1993), 'Fishing town transformed for the weekend', *Nation*, 16 April.

Dann, G.M.S. (1994), 'Tourism: the nostalgia industry of the future', in W. Theobald (ed), *Global Tourism: The Next Decade*, Butterworth-Heinemann, London, pp.55-68.

Dann, G.M.S. and Potter, R.B. (1994), 'Tourism and postmodernity in a Caribbean setting', *Cahiers du Tourisme series C*, no.185 (Universite de Droit, d'Economie et des Sciences, Aix-en-Provence).

Dann, G.M.S. and Potter, R.B. (1997), 'Tourism in Barbados: rejuvenation or decline?', in D. Lockhart and D. Drakakis-Smith (eds), *Island Tourism: Problems and Perspectives*, Mansell, London.

Davies, H. (1993), 'Down to his last shack in the sun', *The Independent*, 25 May, p.16.

de Soto, H. (1989), *The Other Path*, Tauris, London.

Deere, C., Antrobus, P., Bolles, L., Melendez, E., Phillips, P., Rivera, M. and Safa, H. (eds) (1990), *In the Shadows of the Sun: Caribbean Development Alternatives and US Policy*, Westview Press, Colorado.

Demas, W.G. (1965), *The Economies of Development in Small Countries with Special Reference to the Caribbean*, McGill University Press, Montreal.

Denzin, N. (1986), 'Postmodern social theory', *Sociological Theory*, vol.4(2), pp.194-204.

Dicken, P. (1992), *Global Shift*, Harper Row, London and New York.

Dirks, R. (1978), 'Resource fluctuations and competitive transformations in West Indian slave societies', in C.D. Laughlan and I.A. Brady (eds), *Extinction and Survival in Human Populations*, Columbia University Press, New York.

Downes, A.S. (1980), 'A basic needs strategy for the newly independent micro-states', *Bulletin of Eastern Caribbean Affairs*, vol.6, pp.12-22.

Drakakis-Smith, D. (1990), 'Food for thought or thought about food: urban food distribution systems in the Third World', in R.B. Potter and A.T. Salau (eds), *Cities and Development in the Third World*, Mansell, London and New York.

Dufour, R. (1978), 'Des mythes du loisir/tourisme, weekend: aliénation ou libération?', *Cahiers du Tourisme*, série C, no.47.

Durant-Gonzalez, V. (1985), 'Higgerling: rural women in the internal market system in Jamaica', in P.I. Gomes (ed), *Rural Development in the Caribbean*, Hurst & Company, London, pp.103-122.

Eco, U. (1986), *Travels in Hyper-Reality*, Picador, London.

Eliade, M. (1968), *Myths, Dreams and Mysteries*, Fontana, London.

Emanuel, K.A. (1987), 'The dependence of hurricane intensity on climate', *Nature*, vol.326, pp.483-485.

English, E. (1986), *The Great Escape. An examination of north-south tourism*, North-South Institute, Ottawa.

Erisman, H. (1983), 'Tourism and cultural dependency in the West Indies', *Annals of Tourism Research*, vol.10, pp.337-361.

Errington, F. and Gewertz, D. (1989), 'Tourism and anthropology in a post-modern world', *Oceania*, vol.60, pp.37-54.

Farrell, T.M.A. (1978), 'The unemployment crisis in Trinidad and Tobago: its current dimensions and some projections to 1985', *Social and Economic Studies*, vol.27(2), pp.171-185.

Featherstone, M. (1991), *Consumer Culture and Postmodernism*, Sage, London.

Frank, A.G. (1969), *Capitalism and Underdevelopment in Latin America*, Monthly Review Press, New York.

Franklin, G.H. (1979), 'Physical development planning and the Third World', *Third World Planning Review*, vol.1, pp.7-22.

Fraser, P.D. (1985), *Caribbean Economic Handbook*, Euromonitor Publications, London.

Friedmann, J. and Weaver, C. (1979), *Territory and Function: The Evolution of Regional Planning*, Arnold, London.

Frow, J. (1991), *Tourism and the Semiotics of Nostalgia*, October 57, pp.123-151.

Fussell, P. (1979), 'The stationary tourist', *Harpers*, April, pp.31-38.

Gottlieb, A. (1982), 'Americans' vacations', *Annals of Tourism Research*, vol.9, pp.165-187.

Government of Barbados (1970), *Physical Development Plan for Barbados*, Barbados.

Government of Barbados (1983), *Barbados Physical Development Plan Amended 1983*, Barbados.

Government of Barbados (1998), *Draft National Physical Development Plan*, Barbados.

Government of Trinidad and Tobago (1982), *National Physical Development Plan*, Development Planning Series, Republic of Trinidad and Tobago, Port of Spain.

Government of Trinidad and Tobago (1991), *Medium Term Macro Planning Framework 1989-1995*, National Planning Commission, Republic of Trinidad and Tobago, Port of Spain.

Graburn, N. (1992), 'The myth, the real and the hyperreal: a liminal theory of tourism', paper presented to the conference Tourisme International entre Tradition et Modernité, Carrefour Universitaire Méderranéan, Nice, 19-23 November.

Greenwood, D. (1977), 'Culture by the pound: an anthropological perspective on tourism as cultural commoditization', in V. Smith (ed), *Hosts and Guests: The Anthropology of Tourism*, University of Pennsylvania Press, Philadelphia, pp.129-138.

Greiner, T.H. (1977), *Infant Food Advertising and Malnutrition in St Vincent*, MSc Dissertation, Cornell University.

Grenada Planning Office (no date), *Sectoral Issue Papers – Draft Grenada National Physical Development Plan*, St George's, Grenada.

Griffin, E. and Ford, L. (1980), 'A model of Latin American city structure', *Geographical Review*, vol.70, pp.397-422.

Gugler, J. (1980), 'A minimum of urbanism and a maximum of ruralism: the Cuban experience', *International Journal of Urban and Regional Research*, vol.4, pp.516-535.

Gumbs, F. (1981), 'Agriculture in the wider Caribbean', *Ambio*, vol.10, pp.335-339.

Hall, D. (1981), 'Town and country planning in Cuba', *Town and Country Planning*, vol.50, pp.81-83.

Hall, P. (1974), *Urban and Regional Planning*, Penguin, Harmondsworth.

Harrison, F. (1991), 'Women in Jamaica's urban informal economy: insights from a Kingston slum', in C. Mohanty, A. Russo and L. Torres (eds), *Third World Women and the Politics of Feminism*, Indiana University Press, Bloomington, pp.173-196.

Harrison, R. (1984), 'English speaking Caribbean less developed counries (LDCs): growth and development of the manufacturing sector', *Bulletin of Eastern Caribbean Affairs*, vol.9, pp.15-26.

Harvey, D. (1989), *The Condition of Postmodernity*, Blackwell, Oxford.

Hendy, M. (1993), 'Sea-level movements and shoreline change', in G.A. Maul (ed), *Climate Change in the Intra-Americas Sea*, Edward Arnold, London, pp.115-161.

Henfrey, C. (1984), 'Between populism and Leninism: the Grenadian experience', *Latin American Perspectives*, vol.11, pp.15-36.

Hettne, B. (1995), *Development Theory and the Three Worlds* (second edition), Longman, Harlow.

Hope, K.R. (1986), *Urbanization in the Commonwealth Caribbean*, Westview Press, Boulder, Colorado.

Hudson, B. (1986), 'Landscape as resource for national development: a Caribbean view', *Geography*, vol.71, pp.116-121.

Hudson, B. (1989), 'The Commonwealth Eastern Caribbean', in R.B. Potter (ed), *Urbanization, Planning and Development in the Caribbean*, Mansell, London and New York, pp.181-211 and 255-315.

Hudson, J.C. (1969), 'Diffusion in a central place system', *Geographical Analysis*, vol.1, pp.45-58.

Husbands, W. (1986), 'Periphery resort tourism and tourist-resident stress: an example from Barbados', *Leisure Studies*, vol.5, pp.175-188.

Intergovernmental Panel on Climate Change (1990), *Climate Change: The IPCC Scientific Assessment*, Cambridge University Press, Cambridge.

Intergovernmental Panel on Climate Change (1992), *Climate Change 1992: The Supplementary Report to the IPCC Scientific Assessment*, Cambridge University Press, Cambridge.

Ishmael, L. (1988), *Informal Sector Factor Mobilization: the process by which poor people shelter themselves and implications for policy focus on the Caribbean: St Vincent and Dominica*, unpublished PhD thesis, University of Pennsylvania.

Ishmael, L. (1989), *Housing Sector Overview*, Government of St Vincent and the Grenadines and United Nations Center for Human Settlements and United Nations Development Programme.

Jameson, K.P. (1981), 'Socialist Cuba and the intermediate regimes of Jamaica and Guyana', *World Development*, vol.9, pp.871-888.

Jameston, F. (1985), 'Post-modernism and consumer culture', in H. Foster (ed), *Postmodern Culture*, Pluto, London, pp.111-125.

Katzin, M.F. (1960), 'The business of higglering in Jamaica', *Social and Economic Studies*, vol.9, pp.297-331.

Kelly, D. (1986), 'St Lucia's female electronics factory workers: key components in an export-oriented industrialisation strategy', *World Development*, vol.14, pp.823-838.

Kincaid, J. (1988), *A Small Place*, Farrar Strauss Giroux, New York.

Kirton, C. and Witter, M. (1993), 'Aspects of the informal economy in Jamaica', in S. Lalta and M. Frekelton (eds), *Caribbean Economic Development: The First Generation*, Ian Randle, Kingston, pp.280-290.

Klak, T. (ed) (1998), *Globalization and Neoliberalism: The Caribbean Context*, Rowman and Littlefield, Lanham.

Kowalewski, D. (1982), *Transnational Corporations and Caribbean Inequalities*, Praeger, New York.

Lanfant, M-F. (1980), 'Tourism in the process of internationalization', *International Social Science Journal*, vol.XXXII(1), pp.14-43.

Lash, S. (1991), *Sociology of Postmodernism*, Routledge, London.

Latin American and Caribbean Commission on Development and Environment (1990), *Our Own Agenda*, Inter-American Bank and United Nations Development Programme, New York.

Le Franc, E. (1989), 'Petty trading and labour mobility: higglers in the Kingston metropolitan area', in K. Hart (ed), *Women and the Sexual Division of Labour in the Caribbean*, Consortium Graduate School of Social Sciences, University of the West Indies, Kingston, pp.99-132.

Lett, J. (1983), 'Ludic and liminoid aspects of charter yacht tourism in the Caribbean', *Annals of Tourism Research*, vol.10, pp.35-56.

Lewis, W.A. (1950), 'The industrialisation of the British West Indies', *Caribbean Economic Review*, vol.2, pp.1-61.

Lewis, W.A. (1955), *The Theory of Economic Growth*, George Allen & Unwin, London.

Lloyd Evans, S. (1994), *Ethnicity and Gender in the Informal Sector in Trinidad: With Particular Reference to Petty-commodity Trading*, unpublished PhD thesis, University of London.

Lloyd Evans, S. and Potter, R.B. (1992), 'The informal sector of the economy of the Commonwealth Caribbean', *Bulletin of Eastern Caribbean Affairs*, vol.17(3), pp.26-40.

Lloyd, P. (1979), *Slums of Hope? Shanty Towns of the Third World*, Penguin, Harmondsworth.

Lloyd, P. (1999), *Debt and Adjustment: Social and Environmental Consequences in Jamaica*, Ashgate, Aldershot.

Louis, E.L. (1986), *A Critical Analysis of Low-Income Housing in St Lucia*, MSc dissertation, St Augustine Campus of the University of the West Indies.

Lowder, S. (1986), *Inside Third World Cities*, Croom Helm, Beckenham.

Lowenthal, D. (1961), 'Caribbean views of Caribbean land', *Canadian Geographer*, vol.2, pp.1-9.

Lowenthal, D. (1972), *West Indian Societies*, Oxford University Press, London.

Lowenthal, D. (1985), *The Post is a Foreign Country*, Cambridge University Press, Cambridge.

Lowenthal, D. (1987), 'Foreword', in J. Besson and J. Momsen (eds), *Land and Development in the Caribbean*, Macmillan, London.

MacCannell, D. (1976), *The Tourist: A New Theory of the Leisure Class*, Schocken, New York.

MacCannell, D. (1992), *Empty Meeting Grounds: The Tourist Papers*, Routledge, London.

MacLeod, S. and McGee, T. (1990), 'The last frontier: the emergence of the industrial palate in Hong Kong', in D. Drakakis-Smith (ed), *Economic Growth and Urbanization in Developing Areas*, Routledge, London and New York.

McAffe, K. (1991), *Storm Signals: Structural Adjustment and Development Alternatives in the Caribbean*, Zed Books, London.

McElroy, J. and de Albuquerque, K. (1986), 'The tourism demonstration effect in the Caribbean', *Journal of Travel Research*, vol.25, pp.31-34.

McElroy, J. and de Albuquerque, K. (1989), 'Puerto Rico and the United States Virgin Islands', in R.B. Potter (ed), *Urbanization, Planning and Development in the Caribbean*, Mansell, London and New York.

McElroy, J. and de Albuquerque, K. (1989), 'Tourism styles and policy responses in the open-economy closed environment context', paper presented at the

Caribbean Conservation Association conference on economics and the environment, Barbados, 6-8 November.

McElroy, J. and de Albuquerque, K. (1991), 'An integrated sustainable tourism for small islands', paper presented at the XVI Annual Conference of the Caribbean Studies Association, Havana, Cuba.

McElroy, J. and de Albuquerque, K. (1993), 'Sustainable alternatives to insular mass tourism: recent theory and practice', paper presented to the conference on Sustainable Tourism in Islands and Micro-States, Foundation for International Studies, University of Malta, Valetta, 18-20 November.

McFarlane, A. (1989), *The Beach Economy*, Caribbean Tourism Organisation, Christ Church, Barbados.

McGee, T. (1979), 'Conservation and dissolution in the Third World city: the 'shanty town' as an element of conservation', *Development and Change*, vol.10, pp.1-22.

McGregor, D.F.M. and Barker, D. (1991), 'Land degradation and hillside farming in the Fall River Basin, Jamaica', *Applied Geography*, vol.11, pp.143-156.

McGregor, D.F.M. and Potter, R.B. (1997), 'Environmental change and sustainability in the Caribbean: terrestrial perspectives', in B.M. Ratter and W-D. Sahr (eds), *Land, Sea and Human Effort in the Caribbean*, University of Hamburg, Hamburg, pp.1-15.

Miller, R. (1992), 'Tennant's lager', *Sunday Times Magazine*, 1 March, pp.28-36.

Miller, S. and Miller, K. (1991), *Ins and Outs of Barbados*, Miller Publishing Company, Prior Hill, St James, Barbados.

Milliman, J.D. (1993), 'Coral reefs and their response to global climate change', in G.A. Maul (ed), *Climate Change in the Intra-Americas Sea*, Edward Arnold, London, pp.306-321.

Ministry of Housing and Lands (1984), *White Paper on Housing*, Barbados Government Printing Office.

Mintz, S.W. (1985), *Sweetness and Power: The Place of Sugar in Modern History*, Viking, New York.

Moore, A. (1980), 'Walt Disney World: bounded ritual space and the playful pilgrimage center', *Anthropological Quarterly*, vol.53(4), pp.207-218.

Morgan, P. (1992), 'Letter to the Editor', *Barbados Advocate*, 12 December.

Morris, A.S. (1976), 'Urban growth patterns in Latin America with illustrations from Caracas', *Urban Studies*, vo.15, pp.299-312.

Moscardo, G. and Pearce, P. (1986), 'Historic theme parks: an Australian experience in authenticity', *Annals of Tourism Research*, vol.13, pp.467-479.

Mount Gay (1992), *Welcome to the home of Mount Gay*, In House Graphics, Bridgetown.

Nanita-Kennett, M. (1998), 'Industrial free zones in the Dominican Republic', in D. Barker, C. Newby and M. Morrissey (eds), *A Reader in Caribbean Geography*, Ian Randle Publishers, Kingston, pp.101-106.

Nanton, P. (1983), 'The changing pattern of state control in St Vincent and the Grenadines', in F. Ambursley and R. Cohen (eds), *Crisis in the Caribbean*,

Heinemann, London.

Nation (1993), 'Agents sample Bajan-style cooking', 25 November.

Nicholson, C. (1984), *Chase the Moon*, Sphere Books, London.

Nurse, L.A., Atherley, K.A. and Brewster, L.A. (1995), 'The impact of thermal effluent discharge on the Barbados south west coast', in D. Barker and D.F.M. McGregor (eds), *Environment and Development in the Caribbean: Geographical Perspectives*, University of the West Indies Press, Kingston, pp.21-34.

Papson, S. (1979), 'Tourism: world's biggest industry in the twenty-first century', *The Futurist*, vol.XIII(4), pp.249-258.

Patullo, P. (1997), *Last Resort? Tourism in the Caribbean*, Mansell and the Latin America Bureau, London.

Payne, A. (1984), *The International Crisis in the Caribbean*, Croom Helm, London.

Pearson, R. (1993), 'Gender and new technology in the Caribbean: new work for women?', in J.H. Momsen (ed), *Women and Change in the Caribbean: A Pan-Caribbean Perspective*, James Currey, London, pp.287-295.

Pederson, P.O. (1970), 'Innovation diffusion within and between national urban systems', *Geographical Analysis*, vol.2, pp.203-254.

Phillip, M.P. (1988), *Urban Low Income Housing in St Lucia: an analysis of the formal and informal sectors*, unpublished MPhil thesis, University of London.

Phillips, D. (1985), *Women Traders in Trinidad and Tobago*, Economic Commission for Latin America, Port of Spain.

Poon, A. (1993), *Tourism, Technology and Competitive Strategies*, CAB International, Wallingford UK.

Portes, A., Castells, M. and Benton, L.A. (eds) (1989), *The Informal Economy: Studies in Advanced and Less Developed Countries*, John Hopkins University Press, Baltimore.

Portes, A., Dore-Cabral, C. and Landolt, P. (1997), *The Urban Caribbean: Transition to the New Global Economy*, John Hopkins University Press, Baltimore.

Potter, R.B. (1981), 'Industrial development and urban planning in Barbados', *Geography*, vol.66, pp.225-228.

Potter, R.B. (1982), *The Urban Retailing System: Location, Cognition and Behaviour*, Gower, Aldershot.

Potter, R.B. (1983), 'Tourism and development: the case of Barbados, West Indies', *Geography*, vol.68, pp.46-50.

Potter, R.B. (1984), 'Spatial perceptions and public involvement in Third World urban planning: the example of Barbados', *Singapore Journal of Tropical Geography*, vol.5, pp.30-44.

Potter, R.B. (1985), *Urbanisation and Planning in the Third World: Spatial Perceptions and Public Participation*, Croom Helm and St Martin's Press, London and New York.

Potter, R.B. (1986), 'Physical development or spatial land use planning in

Barbados: retrospect and prospect', *Bulletin of Eastern Caribbean Affairs*, vol.12, pp.24-32.

Potter, R.B. (1987), 'Spatial inequalities in Barbados', *Transactions of the Institute of British Geographers, New Series*, vol.11, pp.183-198.

Potter, R.B. (1989a), *Urbanization, Planning and Development in the Caribbean*, Mansell, London.

Potter, R.B. (1989b), 'Urbanization, planning and development in the Caribbean: an introduction', in R.B. Potter (ed), *Urbanization, Planning and Development in the Caribbean*, Mansell, London and New York, pp.1-20.

Potter, R.B. (1989c), 'Rural-urban interaction in Barbados and the Southern Caribbean: patterns and processes of dependent development in small countries', in R.B. Potter and T. Unwin (eds), *The Geography of Urban-Rural Interaction in Developing Countries*, Routledge, London and New York, pp.257-293.

Potter, R.B. (1989d), 'Urban housing in Barbados, West Indies', *Geographical Journal*, vol.155, pp.81-93.

Potter, R.B. (1989e), 'Urban-rural interaction, spatial polarisation and development planning', in R.B. Potter and T. Unwin (eds), *The Geography of Urban-Rural Interaction in Developing Countries*, Routledge, London and New York, pp.323-333.

Potter, R.B. (1990), 'Cities, convergence, divergence and Third World development', in R.B. Potter and A.T. Salau (eds), *Cities and Development in the Third World*, Mansell, London and New York.

Potter, R.B. (1991a), 'A note concerning housing conditions in Grenada, St Lucia and St Vincent', *Bulletin of Eastern Caribbean Affairs*, vol.16, pp.13-23.

Potter, R.B. (1991b), 'An analysis of housing in Grenada, St Lucia and St Vincent and the Grendines', *Caribbean Geography*, vol.3, pp.106-125.

Potter, R.B. (1991c), 'Caribbean popular housing: Cinderella of research and policy', *Caribbean Studies Newsletter*, vol.18, pp.19-21.

Potter, R.B. (1992a), *Housing Conditions in Barbados: A Geographical Analysis*, University of the West Indies, Barabdos, Jamaica, Trinidad and Tobago.

Potter, R.B. (1992b), 'Demographic change in a small island state: St Vincent and the Grenadines', *Geography*, vol.77(4), pp.381-383.

Potter, R.B. (1992c), 'Caribbean views of Caribbean environment and development', *Caribbean Geography*, vol.3, pp.236-243.

Potter, R.B. (1992d), *Urbanisation in the Third World*, Oxford University Press, Oxford.

Potter, R.B. (1993a), 'Basic need and development in the small island states of the eastern Caribbean', in D. Lockhart, D. Drakakis-Smith and J. Schembri (eds), *The Development Process in Small Island States*, Routledge, London and New York, pp.92-116.

Potter, R.B. (1993b), 'Demographic change in a small island state: St Vincent and the Grenadines', *Geography*, vol.77, pp.374-375.

Potter, R.B. (1993c), 'Urbanization in the Caribbean and trends of global convergence-divergence', *Geographical Journal*, vol.159, pp.1-21.

Potter, R.B. (1995a), 'Urbanisation and development in the Caribbean', *Geography*, vol.80, pp.334-341.

Potter, R.B. (1995b), *Low-Income Housing and the State in the Eastern Caribbean*, University of the West Indies, Mona, Jamaica.

Potter, R.B. and Binns, J.A. (1988), 'Power, politics and society', in M. Pacione (ed), *The Geography of the Third World: Progress and Prospects*, Routledge, London and New York, pp.271-310.

Potter, R.B. and Conway, D. (eds) (1997), *Self-Help Housing, the Poor, and the state in the Caribbean*, Tennessee University Press and The Press University of the West Indies, Knoxville.

Potter, R.B. and Coshall, J. (1984), 'The hand analysis of repertory grids: an appropriate technique for Third World environmental studies', *Area*, vol.16, pp.315-322.

Potter, R.B. and Dann, G.M.S. (1987), 'Introduction', *Barbados: World Bibliographical Series*, Clio Press, Oxford.

Potter, R.B. and Dann, G.M.S. (1987), *World Bibliographical Series Volume 76: Barbados*, Clio Press, Oxford.

Potter, R.B. and Dann, G.M.S. (1990), 'Dependent urbanization and retail change in Barbados, West Indies', in R.B. Potter and A.T. Salau (eds), *Cities and Development in the Third World*, Mansell, London and New York, pp.172-192.

Potter, R.B. and Dann, G.M.S. (1994), 'Some observations concerning postmodernity and sustainable development in the Caribbean', *Caribbean Geography*, vol.5, pp.92-101.

Potter, R.B. and Hunte, M.L. (1979), 'Recent developments in planning the settlement hierarchy of Barbados: implications concerning the debate on urban primacy', *Geoforum*, vol.10, pp.355-362.

Potter, R.B. and Lloyd-Evans, S. (1998), *The City in the Developing World*, Longman, Harlow.

Potter, R.B. and O'Flaherty, P. (1995), 'An analysis of housing conditions in Trinidad and Tobago', *Social and Economic Studies*, vol.44, pp.165-183.

Potter, R.B. and Unwin, T. (1992), 'Urban-rural interaction: physical form and political process in the Third World', *Cities*, vol.12, pp.67-73.

Potter, R.B. and Unwin, T. (eds) (1989), *The Geography of Urban-Rural Interaction in Developing Countries*, Routledge, London and New York.

Potter, R.B. and Welch, B. (1994), 'Indigenization and development in the eastern Caribbean: reflections on culture, diet and agriculture', *Caribbean Geography*, in press.

Potter, R.B. and Wilson, M. (1989), 'Barbados', in R.B. Potter (ed), *Urbanization, Planning and Development in the Caribbean*, Mansell, London and New York.

Potter, R.B. and Wilson, M. (1991), 'Indigenous environmental learning in a small developing country: adolescents in Barbados, West Indies', *Singapore Journal of Tropical Geography*, vol.11, pp.56-67.

Potter, R.B., Binns, J.A., Elliott, J. and Smith, D. (1999), *Geographies of*

Development, Longman, Harlow.

Potter, R.B., Jacyno, J. and Lloyd, M. (1995), 'Socio-economic, demographic and residential conditions in Barbados: a preliminary analysis of the 1990 Census at the parish level', Centre for Developing Areas Research (CEDAR) Paper, no.15, 26pp.

Pred, A. (1977), *City-Systems in Advanced Economies*, Hutchinson, London.

Pugh, J. and Potter, R.B. (1999), 'An overview and critique of the Third National Physical Development Plan for Barbados', *CEDAR Research Papers*, no.28, Royal Holloway, University of London.

Rampersad, M. (1991), 'The measurement of the informal sector in Trinidad and Tobago', Central Statistical Office Research Papers, Port of Spain.

Reading, A.J. and Walsh, R.P.D. (1995), 'Tropical cyclone activity within the Caribbean Basin since 1500', in D. Barker and D.F.M. McGregor (eds), *Environment and Development in the Caribbean: Geographical Perspectives*, University of the West Indies Press, Kingston, pp.124-146.

Relph, E. (1983), *Place and Placelessness*, Pion, London.

Richardson, B. (1992), *The Caribbean in the Wider World 1492-1992*, Cambridge University Press, Cambridge.

Rivers, P. (1972), *The Restless Generation: A City in Mobility*, Davis-Poynter, London.

Robertson, R.E.A. (1987), *Disaster Management in St Vincent and the Grenadines*, BSc dissertation, University of the West Indies.

Rojas, E. (1984), 'Agricultural land in the Eastern Caribbean: from resources for survival to resources for development', *Land Use Policy*, vol.1, pp.39-54.

Rojas, E. (1989), 'Human settlements of the Eastern Caribbean: development problems and policy options', *Cities*, vol.6, pp.243-258.

Rojas, E. and Meganck, R.A. (1987), 'Land distribution and land development in the Eastern Caribbean', *Land Use Policy*, vol.4, pp.157-167.

Safa, H. (1986), 'Economic autonomy and sexual equality in Caribbean society', *Social and Economic Studies*, vol.35(3), pp.1-22.

Safa, H. (1990), 'Women and industrialisation in the Caribbean', in S. Stitcher and J. Parpart (eds), *Women, Employment and the Family in the International Division of Labour*, Macmillan, London, pp.72-97.

Sahr, W-D. (1998), 'Micro-metropolis in the Eastern Caribbean: the example of St Lucia', in D.F.M. McGregor and S. Lloyd-Evans (eds), *Resources, Sustainability and Caribbean Development*, The Press University of the West Indies, Barbados, Jamaica and Trinidad and Tobago.

Santos, M. (1979), *The Shared Space: The Two Circuits of the Urban Economy in Underdeveloped Countries*, Methuen, London.

Scotland, L.H. (1983), 'The petroleum windfall, government activity and economic performance in Trinidad and Tobago', paper presented at the First Annual Conference on Trinidad and Tobago Economy, Port of Spain.

Segre, R. (1991), 'Los valores culturales de la cuidad caribeña: el rescate de su significacíon social', *Caribbean Studies Newsletter*, vol.18, pp.17-19.

Selwyn, T. (1992a), 'Peter Pan in south-east Asia. Views from the brochures', in M. Hitchcock, V. King and M. Parnwell (eds), *Tourism in South-East Asia*, Routledge, London, pp.117-137.

Selwyn, T. (1992b), 'Chasing myths', paper presented to the conference Tourisme Entre Tradition et Modernité, Carrefour Universitaire Méditerranéan, Nice, 19-23 November.

Senior, O. (1991), *Working Miracles: Women's Lives in the English-speaking Caribbean*, James Currey, London.

Shapiro, L.J. (1982), 'Hurricane climatic fluctuations: Part II: Relation to large-scale circulation', *Monthly Weather Review*, vol.110, pp.1014-1023.Wigley, T.M.L. and Santer, B.D. (1993), 'Future climate of the Gulf/Caribbean Basin from the global circulation models', in G.A. Maul (ed), *Climate Change in the Intra-Americas Sea*, Edward Arnold, London, pp.31-54.

Shephard, C.Y. (1947), 'Peasant agriculture in the Leeward and Windward Islands', *Tropical Agriculture*, vol.24, pp.61-71.

Sheridan, R. (1973), *Sugar and Slavery: An Economic History of the British West Indies 1623-1775*, John Hopkins University Press, Baltimore.

Singh, S. (1988), *Housing in the English-Speaking Caribbean: recent conditions and trends*, Central Statistical Printing Office, Trinidad and Tobago.

Slater, D. (1986), 'Socialism, democracy and the territorial imperative: elements for a comparison of the Cuban and Nicaraguan experiences', *Antipode*, vol.18, pp.155-185.

Smithson, P. (1993), 'Tropical cyclones and their changing impact', *Geography*, vol.78, pp.170-174.

Spencer, F. (1974), *Crop Over: An Old Barbadian Plantation Festival*, Commonwealth Caribbean Resource Centre, Barbados.

Stevens, P.H.M. (1957), 'Planning in the West Indies', *Town and Country Planning*, vol.25, pp.503-508.

Sunday Advocate (1992), 'Bajan culture up for sale – Greaves', 13 December.

Sunday Advocate (1993a), 'Community festivals a good thing – Barker-Welch', 11 April.

Sunday Advocate (1993b), 'Rent-a-dread', 18 April.

Sunday Advocate (1993c), 'New Coca Cola campaign begins', 9 May.

Sunday Advocate (1993d), 'Treat skills a business', 23 May.

Sunday Advocate (1993e), 'Barbados Crop Over advertisement', 8 August.

Sunday Advocate (1993f), 'Christmas wonderland at Simpsons', 5 December.

Sunday Sun (1993), 'Advertisement for 1993 Heritage Festival', 28 November.

Susman, P. (1974), 'Cuban development: from dualism to integration', *Antipode*, vol.6, pp.10-29.

Susman, P. (1987), 'Spatial equality and socialist transformation in Cuba', in D. Forbes and N. Thrift (eds), *The Socialist Third World: Urban Development and Territorial Planning*, Blackwell, Oxford, pp.250-281.

Sylvain, R. (1993), 'Heritage passport opens door to Bajan culture', *Toronto Star*, 16 January.

Taffee, E.J., Morrill, R.L. and Gould, P.R. (1963), 'Transport expansion in underdeveloped countries: a comparative analysis', *Geographical Review*, vol.53, pp.503-529.

The Caribbean Centre for Development Administration, The Caribbean Development Bank and the United Nations Development Programme (1997), *Capacity 21 Project: Report of Regional Consultation - Sustainable Development Councils Barbados, the British Virgin Islands, Commonwealth of Dominica, Grenada, Jamaica and St Lucia*, Caribbean Centre for Development Administration, Barbados.

Thomas, C.Y. (1974), *Dependence and Transformation: the Economics of the Transition to Socialism*, Monthly Review Press, New York and London.

Thomas, C.Y. (1988), *The Poor and the Powerless: Economic Policy and Change in the Caribbean*, Latin American Bureau, London.

Thomas, G.A. (1991), 'The gentrification of paradise: St John's, Antigua', *Urban Geography*, vol.12, pp.469-487.

Thomson, R. (1987), *Green Gold: Bananas and Dependency in the Eastern Caribbean*, Latin American Bureau, London.

Thurot, J. (1989), 'Psychologie du loisir touristique', *Cahiers du Tourisme*, série C, no.23.

Toffler, A. (1970), *Future Shock*, Random House, New York.

Toffler, A. (1980), *The Third Wave*, William Morrow, New York.

Tokman, V. (1989), 'Policies for a heterogeneous informal sector in Latin America', *World Development*, vol.17, pp.1067-1076.

Travel and Leisure (1990), 'Play the Bajan way (advertisment)', November, p.251.

Tresse, P. (1990), 'L'image des civilisations africaines à travers les publications des services officiels du tourisme des pays d'Afrique Francophone', *Cahiers du Tourisme*, série C, no.11.

Turner, J.F.C. (1967), 'Barriers and channels for housing development in modernizing countries', *Journal of the American Institute of Planners*, vol.33, pp.167-181.

Turner, J.F.C. (1968), 'Housing priorities, settlement patterns and urban development in modernizing countries', *Journal of the American Institute of Planners*, vol.34, pp.354-363.

Turner, J.F.C. (1976), *Housing by People: Towards Autonomy in Building Environments*, Marion Boyars, London.

Turner, J.F.C. (1982), 'Issues in self-help and self-managed housing', in P. Ward (ed), *Self-Help Housing: A Critique*, Mansell, London.

Turner, L. (1976), 'The international division of leisure. Tourism and the Third World', *Annals of Tourism Research*, vol.4(1), pp.12-24.

Turner, L. and Ash, J. (1975), *The Golden Hordes. International Tourism and the Pleasure Periphery*, Constable, London.

United Nations (1980), *Patterns of Urban and Rural Population Growth*, United Nations, New York.

United Nations (1989), *Prospects for World Urbanization 1988*, United Nations,

New York.

United Nations (1994), *Earth Summit: Programme of Action for Small Island States*, Global Conference on the Sustainable Development of Small Island Developing States, Bridgetown, Barbados, 26 April to 6 May 1994, United Nations, New York.

Urry, J. (1990), *The Tourist Gaze*, Sage, London.

Urry, J. (1991), 'Tourism, travel and the modern subject', *Vr ijetijd en Samenleving*, vol.9(3/4), pp.87-98.

Vance, J.E. (1970), *The Merchant's World: The Geography of Wholesaling*, Prentice-Hall, Englewood Cliffs.

Vendovato, C. (1986), *Politics, Foreign Trade and Economic Development: A Study of the Dominican Republic*, Croom Helm, London.

Visitor Magazine (1993a), 29 March to 4 April.

Visitor Magazine (1993b), 3-9 May.

Wagner, U. (1977), 'Out of time and place – mass tourism and charter trips', *Ethnos*, vol.42, pp.38-52.

Wallace, I. (1990), *The Global Economic System*, Unwin Hyman, London.

Watts, D. (1987), *The West Indies: Patterns of Development, Culture and Environmental Change since 1492*, Cambridge University Press, Cambridge.

West, R.C. and Augelli, J.P. (1976), *Middle America: Its Land and Peoples*, Prentice-Hall, Englewood Cliffs, New Jersey.

Widfelt, A. (1993), 'The impact of alternative development strategies on tourism in Caribbean microstates', paper presented to the conference on Sustainable Tourism in Islands and Micro-States, Foundation for International Studies, University of Malta, Valetta, 18-20 November.

Wigley, T.M.L. and Santer, B.D. (1993), 'Future climate of the Gulf/Caribbean Basin from the global circulation models', in G.A. Maul (ed), *Climate Change in the Intra-Americas Sea*, Arnold, London, pp.31-54.

Wilson, D. (1993), 'Glimpses of Caribbean tourism and the question of sustainability in Barbados and St Lucia', paper presented to the conference on Sustainable Tourism in Islands and Micro-States, Foundation for International Studies, University of Malta, Valetta, 18-20 November.

Wirt, D.S. (1987), 'An assessment of socio-economic conditions of human settlements', Department of Regional Development, Organization of American States, Grenada.

Wolf, E.R. (1982), *Europe and the People without History*, University of California Press, Berkeley.

Wolpe, H. (ed) (1980), *The Articulation of Modes of Production*, Routledge and Kegan Paul, London.

Worrell, D. (1987), *Small Island Economies: Structure and Performance in the English-speaking Caribbean since 1970*, Praeger, New York.

Zetter, R. (1981), 'Imported or indigenous planning education? Some observations on the needs of developing countries', *Third World Planning Review*, vol.3, pp.24-42.

Index

For Product Safety Concerns and Information please contact our EU
representative GPSR@taylorandfrancis.com Taylor & Francis Verlag GmbH,
Kaufingerstraße 24, 80331 München, Germany

Printed and bound by CPI Group (UK) Ltd, Croydon, CR0 4YY

01/05/2025

01858578-0001